Ernst Schering Research Foundation Workshop 29
Of Fish, Fly, Worm, and Man
Lessons from Developmental Biology for Human Gene Function
and Disease

Springer-Verlag Berlin Heidelberg GmbH

Ernst Schering Research Foundation
Workshop 29

Of Fish, Fly, Worm, and Man

Lessons from Developmental Biology
for Human Gene Function and Disease

C. Nüsslein-Volhard, J. Krätzschmar
Editors

With 36 Figures and 5 Tables

 Springer

Series Editors: G. Stock and M. Lessl

ISSN 0947-6075
ISBN 978-3-662-04266-3

CIP data applied for

Die Deutsche Bibliothek – CIP-Einheitsaufnahme
Schering-Forschungsgesellschaft <Berlin>: Ernst Schering Research Foundation Workshop.

ISSN 0947-6075
29. Of fish, fly, worm and man. - 2000
Of fish, fly, worm and man: lessons from developmental biology for human gene function and disease ; with tables / Ch. Nüsslein-Vollhard and J. Krätzschmar, ed.
(Ernst Schering Research Foundation Workshop; 29)
ISBN 978-3-662-04266-3   ISBN 978-3-662-04264-9 (eBook)
DOI 10.1007/978-3-662-04264-9

© Springer-Verlag Berlin Heidelberg 2000
Originally published by Springer-Verlag Berlin Heidelberg New York in 2000
Softcover reprint of the hardcover 1st edition 2000

Typesetting: Data conversion by Springer-Verlag

SPIN:10691251   21/3134/AG–5 4 3 2 1 0 – Printed on acid-free paper

Preface

The first complete genome sequence of a multicellular organism, *Caenorhabditis elegans,* has been determined recently. Several more will follow soon, among them the genome sequences of *Drosophila melanogaster* and the human. With these achievements, the stage is set for the next phase of "large-scale" biology, the study of the function of genes and the interactions between their protein products. The aim of the Ernst Schering Research Foundation Workshop 29 was to discuss the contribution of research on model organisms which are particularly suited for genetic and biological research, to the understanding of general principles of biology as well as the basis for human diseases. This area of research has not only unveiled the basic molecular mechanisms of development of higher organisms, but has also made major contributions to the elucidation of complex signal transduction pathways that play crucial roles both in ontogeny and human diseases.

Establishing the complete molecular anatomy of an organism is a demanding but rather well-defined task. However, compared to the technical and scientific challenges of large-scale sequencing and the bioinformatics analysis of complete genome data, the assignment of function to the genes of multicellular organisms may still seem like a fishing expedition, with no universal road to success.

One approach is based on large-scale technologies. The use of microarrays for multiple gene expression analysis is one element in this effort, providing important information on the cell- or tissue-specific activation of genes over time. Mass spectrometry combined with high-resolution two-dimensional electrophoresis has enabled the systematic study of the protein inventories of cells and tissues, complementing

The organisers and speakers of the workshop

and adding to the information obtained from the study of mRNAs. In addition, systematic analyses of protein–protein interactions using two-hybrid technologies are conducted in the hope of discovering functional interactions. However, in order to draw a more complete picture than is apparent from such descriptive data, it is desirable to study the in vivo function of genes and their protein products in a direct fashion.

To this end, genetic studies in model organisms offer a number of practical advantages, such as short generation times, smaller number of genes and probably fewer functional redundancies within protein families, smaller genomes, and the possibility to conduct saturation mutagenesis screens. The comparative analyses of genes of higher organisms have uncovered a surprisingly high degree of similarities not only in the structures, but also the functional contexts of many genes. This means that the study of nematodes, insects, and zebrafish can provide information that is of great relevance to the discovery of fundamental biological mechanisms at work in higher vertebrates and to the elucidation of mammalian gene function, ultimately leading to the understanding of human pathogenesis.

Genetic approaches have been very successful in identifying proteins functionally implicated in a specific process, through the isolation and characterization of mutations causing certain phenotypes. The general strategy is not so much to define the function of a given gene, but rather, in the absence of any information on the molecules and signaling pathways involved in a particular aspect of development, to identify multiple genes, known or novel, by virtue of a common phenotype. The concept that a single signaling pathway usually plays several roles in both development and adult physiology is supported by a number of studies, in which pathways initially implicated in development, when deregulated, were also found to contribute to the pathogenesis of, for example, colon carcinoma, medulloblastoma, and basal cell carcinoma of the skin. Therefore, developmental biology teaches us not only about mechanisms of ontogenesis but also about the components of, and their functional interplay in, signaling pathways involved in a wide range of intra- and inter-cellular processes.

The primary intent of the Ernst Schering Research Foundation Workshop 29, held in Berlin on January 27–29, 1999, was to review the concepts, methods, and exemplary achievements of genetic research on selected model organisms from the perspective of biomedical research. Examples are provided in the contributions on research in nematodes such as *C. elegans*, concerning programmed cell death and neurodegenerative diseases. Signaling pathways are generally highly conserved between organisms, and several studies were reviewed for *Drosophila*, *Xenopus*, and rodents. As a comparatively novel model organism, the zebrafish was presented in its role of helping to understand cardiovascular functions in surprising detail. Seeing the enthusiasm, talent, and impressive achievements displayed throughout the presentations was an enlightening experience. However, there was also a note of sadness, as one of the speakers originally scheduled to present his work in zebrafish, Dr. Nigel Holder, had died only weeks before the workshop, on December 11, 1998. His manuscript appears in this book, along with a tribute by Dr. Stephen Wilson of the University College London.

Christiane Nüsslein-Volhard
Jörn Krätzschmar

Table of Contents

List of Editors and Contributors

Editors

C. Nüsslein-Volhard
Max-Planck-Institut für Entwicklungsbiologie, Abteilung Genetik,
Spemannstrasse 35/III, 72076 Tübingen, Germany

J. Krätzschmar
Enabling Technologies, Preclinical Drug Research, Schering AG,
Müllerstrasse 178, 13342 Berlin, Germany

Contributors

M. Affolter
Abteilung Zellbiologie, Biozentrum Universität Basel, Klingelbergstrasse 70,
4053 Basel, Switzerland

E.C. Bailey
Department of Cell Biology, MCLM 310, University of Alabama at Birmingham, 1918 University Boulevard, Birmingham, AL 35294-0005, USA

H. Bauer
Max-Planck-Institut für Immunbiologie, Stübeweg 51, 79108 Freiburg,
Germany

T. Bouwmeester
EMBL, Meyerhofstrasse 1, 69117 Heidelberg, Germany

M. Brink
Howard Hughes Medical Institute, Department of Developmental Biology,
Beckman Center, Stanford University Medical School,
Stanford, CA 94305-5323, USA

K. Cadigan
Howard Hughes Medical Institute, Department of Developmental Biology,
Beckman Center, Stanford University Medical School,
Stanford, CA 94305-5323, USA

J.-N. Chen
Cardiovascular Research Center, Massachusetts General Hospital,
149 13th Street, 4th floor, Charlestown, MA 02129-2060, USA

B. Conradt
Max-Planck-Institut für Neurobiologie, Am Klopferspitz 18A,
82152 Martinsried, Germany

J. Cooke
Department of Anatomy and Developmental Biology, University College,
Gower Street, London, WC1 6BT, UK

A. Dick
Max-Planck-Institut für Immunbiologie, Stübeweg 51, 79108 Freiburg,
Germany

R. Dosch
Division of Molecular Embryology, Deutsches Krebsforschungszentrum,
Im Neuenheimer Feld 280, 69120 Heidelberg, Germany

U. Drescher
Max-Planck-Institut für Entwicklungsbiologie, Spemannstrasse 35,
72076 Tübingen, Germany

L. Durbin
Department of Anatomy and Developmental Biology, University College,
Gower Street, London, WC1 6BT, UK

R. Durbin
The Sanger Centre, Wellcome Trust Genome Campus,
Hinxton, Cambs CB10 1SA, UK

M. Fish
Howard Hughes Medical Institute, Department of Developmental Biology,
Beckman Center, Stanford University Medical School,
Stanford, CA 94305-5323, USA

M. C. Fishman
Cardiovascular Research Center, Massachusetts General Hospital,
149 13th Street, 4th floor, Charlestown, MA 02129-2060, USA

C. Haass
Central Institute of Mental Health, Department of Molecular Biology, J5,
68159 Mannheim, Germany

P. Haffter
Max-Planck-Institut für Entwicklungsbiologie, Spemannstrasse 35/III,
72076 Tübingen, Germany

M. Hammerschmidt
Speemann-Laboratorium, Max-Planck-Institut für Immunbiologie,
Stübeweg 51, 79108 Freiburg, Germany

C. Harryman-Samos
Howard Hughes Medical Institute, Department of Developmental Biology,
Beckman Center, Stanford University Medical School,
Stanford, CA 94305-5323, USA

E. Hazendonk
Division of Molecular Biology, The Netherlands Cancer Institute,
Center for Biomedical Genetics, Plesmanlaan 121, 1066 CX Amsterdam,
The Netherlands

M. Hild
Max-Planck-Institut für Immunbiologie, Stübeweg 51, 79108 Freiburg,
Germany

N. Holder†
formerly: Department of Anatomy and Developmental Biology,
University College, Gower Street, London, WC1 6BT, UK

M. van der Horst
Division of Molecular Biology, The Netherlands Cancer Institute,
Center for Biomedical Genetics, Plesmanlaan 121, 1066 CX Amsterdam,
The Netherlands

G. Jansen
Division of Molecular Biology, The Netherlands Cancer Institute,
Center for Biomedical Genetics, Plesmanlaan 121, 1066 CX Amsterdam,
The Netherlands

R.L. Johnson
Department of Cell Biology, MCLM 310, University of Alabama at Birmingham, 1918 University Boulevard, Birmingham, AL 35294-0005, USA

C. Niehrs
Division of Molecular Embryology, Deutsches Krebsforschungszentrum,
Im Neuenheimer Feld 280, 69120 Heidelberg, Germany

R. Nusse
Howard Hughes Medical Institute, Department of Developmental Biology,
Beckman Center, Stanford University Medical School,
Stanford, CA 94305-5323, USA

D. Onichtchouk
Division of Molecular Embryology, Deutsches Krebsforschungszentrum,
Im Neuenheimer Feld 280, 69120 Heidelberg, Germany

R.H.A. Plasterk
Division of Molecular Biology, The Netherlands Cancer Institute,
Center for Biomedical Genetics, Plesmanlaan 121, 1066 CX Amsterdam,
The Netherlands

E. Rulifson
Howard Hughes Medical Institute, Department of Developmental Biology,
Beckman Center, Stanford University Medical School,
Stanford, CA 94305-5323, USA

S. Schulte-Merker
Max-Planck-Institut für Entwicklungsbiologie, Spemannstrasse 35/III,
72076 Tübingen, Germany

M.P. Scott
Departments of Developmental Biology and Genetics,
Howard Hughes Medical Institute, Stanford University School of Medicine,
Stanford, CA 94305-5427, USA

K.L. Thijssen
Division of Molecular Biology, The Netherlands Cancer Institute,
Center for Biomedical Genetics, Plesmanlaan 121, 1066 CX Amsterdam,
The Netherlands

A. Vortkamp
Max-Planck-Institut für Molekulare Genetik, Otto-Warburg-Laboratorium,
Ihnestrasse 73, 14195 Berlin, Germany

P. Werner
Synthélabo Biomoléculaire, 16 Rue d'Ankara, 67080 Strasbourg, France

S. Wilson
Department of Anatomy and Developmental Biology,
University College, Gower Street, London, WC1 6BT, UK

C-h. Wu
Howard Hughes Medical Institute, Department of Developmental Biology,
Beckman Center, Stanford University Medical School,
Stanford, CA 94305-5323, USA

1 Interactions Between Wingless and Frizzled Molecules in Drosophila

R. Nusse, E. Rulifson, M. Fish, C. Harryman-Samos, M. Brink, C-h. Wu, and K. Cadigan

1.1 Introduction

Over the past decade, our understanding of the mechanisms underlying growth and patterning of tissues has been enormously increased due to a convergence of the fields of developmental genetics and cancer research. Systematic screens in organisms such as *Drosophila*, *Caenorhabditis elegans,* and more recently the zebrafish have produced a wealth of genes and proteins that regulate development. At the same time, searches for genes implicated in cancer, either as dominant or as recessive (tumor suppressor) genes, have yielded an equally impressive list of important regulators of growth. There are now also numerous examples of genes found to control normal growth but to cause cancerous growth when misregulated. One of the best examples is the Wnt gene family. We will summarize our current view of how Wnt proteins signal, and the evidence that proteins of the Frizzled (Fz) family mediate Wnt signaling by acting as specific cell surface receptors.

1.2 Wingless in *Drosophila* and Other Wnt Genes

Wnt genes are implicated in numerous developmental processes, as well as in cancer. The *Drosophila* Wnt gene wingless (*wg*) controls many developmental steps and is important for the growth and patterning of tissues, for example the imaginal discs. The *Drosophila* genome contains three other Wnts, as far as is known. All members of the Wnt family encode secreted signaling proteins with one or several N-linked glycosylation sites and many cysteine residues (reviewed in Cadigan and Nusse 1997).

A key step in Wg signaling is mediated by the Armadillo (Arm) protein. Arm is a homolog of the cell junction protein β-catenin but it can also form a complex with a TCF protein. This complex is thought to activate transcription of downstream genes. (Behrens et al. 1996; Miller and Moon 1996; Molenaar et al. 1996). Mutant forms of Arm, as found in several human tumors, can act as constitutively active transcriptional activators (reviewed in Nusse 1997; Peifer 1997).

1.3 In Vitro Assays for Wg Protein and Receptor Identification

The genetic screens in *Drosophila* did not yield a candidate gene for a specific cell surface receptor for Wg; a void now explained by genetic redundancy (see below). Suitable cell culture assays for signaling by Wg and other Wnt proteins have been notoriously difficult, because of problems in handling Wnt proteins in vitro, complicating receptor identification by biochemical tools. It was therefore important that we found that Wg protein is present in soluble form in the medium of transfected *Drosophila* S2 cells and elevates the concentration of the Arm protein in another cell line, clone-8 derived from *Drosophila* imaginal discs (Van Leeuwen et al. 1994).

S2 cells themselves are unable to respond to Wg (Yanagawa et al. 1995). This observation proved to be the key in identifying Wg receptors, as we found that a *Drosophila* frizzled-related gene, Dfz2, identified in the laboratory of our collaborator Dr. Jeremy Nathans, is expressed in clone-8 cells but not in nonresponding S2 cells. When we transfected the Dfz2 gene into S2 cells, these cells became active in Wg signal transduction. In addition, the S2 cells could bind Wg protein on their cell surface. Transfection of cells with Dfz2 constructs lacking either the extracellular or intracellular domain of the protein demonstrated that the extracellular domain was required for binding (Bhanot et al. 1996). We tested several other members of the Fz family in the same assays, finding that the original Fz protein, which genetically had not been implicated in Wg signaling, can confer *wg* responsiveness to S2 cells, as well as Wg binding. A more distant relative, the Smoothened (Smo) protein, does not bind, showing that these assays can discriminate between family members.

These in vitro results suggest that Dfz2 and Fz could have overlapping functions, a hypothesis that we could strengthen recently from genetic data described below. Although these findings solved a long-standing problem in the field, we could not provide evidence that Dfz2 is required for Wg signaling in the fly, because there were no mutants in the gene.

1.4 Wg Signaling

These and other experiments have led to the following current model of Wnt/wg signaling, which is based on both genetics in *Drosophila* and cell biological data (Fig. 1). In the absence of Wg signaling, the protein kinase Zw3 (or GSK-3) inactivates Arm, possibly by direct phosphorylation of Arm followed by proteolytic breakdown. This effect of Zw3 on Arm is then relieved by Wg, resulting in stabilization and upregulation of the Arm protein. The proteins Axin and APC are both involved in downregulating Arm, by forming a complex with Arm and Zw3. After Arm has escaped from these negative regulators, it can act in the nucleus

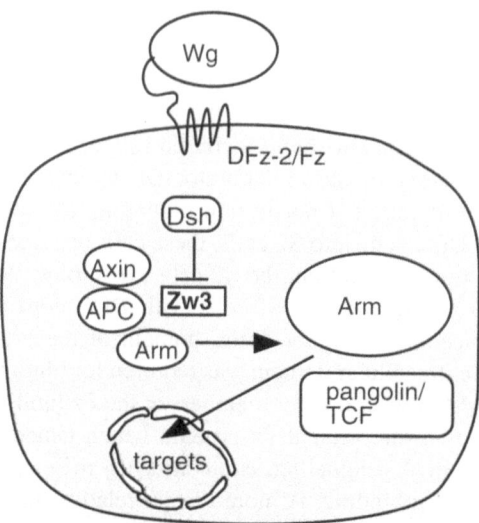

Fig. 1. Wnt/Wg signaling

together with TCF to activate target genes (reviewed in Wodarz and Nusse 1998).

1.5 Fz Molecules and Planar Polarity

fz and a variety of other genes, including *dishevelled* (*dsh*), belong to a class of mutants with planar polarity (PP) phenotypes (Adler 1992; Gubb and Garcia Bellido1982; Shulman et al. 1998). Loss of function for these genes gives a similar phenotype of disoriented bristles and hairs in the adult epidermis; these defects reflect PP defects in the cytoskeleton of the epithelial cells that give rise to the overlying cuticular structure (Gubb and Garcia Bellido 1982). In mosaic clones, *fz* mutations display both cell-autonomous and nonautonomous phenotypes, supporting the role of Fz protein in cell signaling (Vinson and Adler 1987), and suggesting that it interacts with an extracellular ligand. The nature of this ligand during PP signaling is not known, however. While some overexpression studies have left open the possibility that *wg*

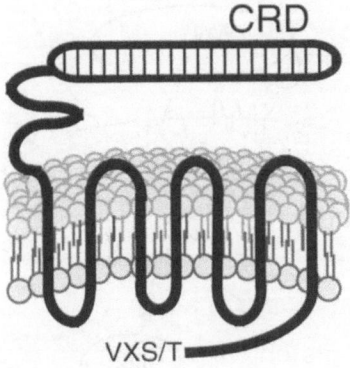

Fig. 2. Structure of Frizzled proteins

acts on *fz*, most studies have not revealed any role for *wg* in PP, suggesting another ligand, possibly another *Drosophila* Wnt.

Fz proteins are part of the large family of seven membrane-spanning domain receptors (Fig. 2). In *Drosophila*, this gene family counts several members, including Dfz2 (Bhanot et al. 1996). The Smoothened protein (Smo), known by genetic analysis to be a component of Hedgehog signaling, is also homologous to the Fz family, yet is the most divergent family member at the sequence level (Alcedo et al. 1996; Van Den Heuvel and Ingham 1996).

Fz proteins contain several structural similarities: a conserved extracellular region containing ten invariant cysteines (CRD); seven hydrophobic segments; and a C-terminus containing a valine residue in a somewhat conserved threonine/serine-X-valine motif (T/SXV), a putative binding site for the PDZ domain (Gomperts 1996; Wang et al. 1996).

Little is known about signaling mechanisms by these receptors. Although the Fz proteins show no primary sequence homology to other known proteins, the structural motifs are reminiscent of G-protein coupled seven-transmembrane segment receptors, with extracellular N-terminal ligand-binding domains, and cytosolic C-termini (Wang et al. 1996). These results suggest a similar structure for Fz proteins and G-protein coupled receptors, although with many G-protein coupled receptors, residues in the transmembrane domains or extracellular loops

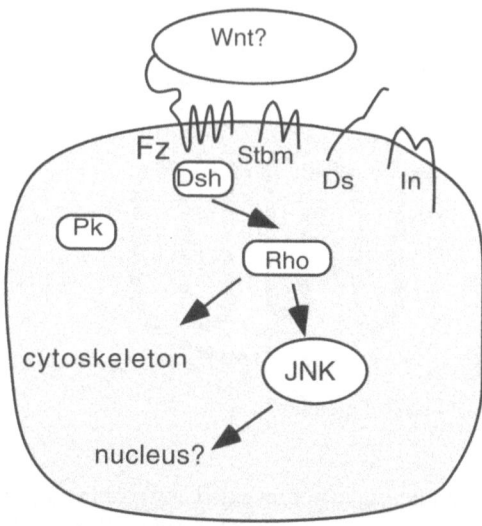

Fig. 3. Fz signaling

are required for ligand binding (reviewed in Dohlman et al. 1991), whereas the CRD of Fz molecules is implicated in ligand binding (see below).

With regard to effector binding, the Fz proteins lack the most conserved motifs present in almost all G-protein coupled receptors, such as the Glu/Asp-Arg-Tyr/Phe triplet at the C-terminus of the third transmembrane segment (Probst et al. 1992; Wang et al. 1996). This suggests that Fz proteins may use novel downstream effector pathways, although there is evidence that at least one Fz member can signal through calcium release and a G-protein (Slusarski et al. 1997). A further genetic analysis of this pathway in organisms such as *Drosophila* and *C. elegans* will undoubtedly shed more light on this issue.

1.6 Signaling Downstream of Fz

In genetic tests, *dsh*, which is also in the Wg pathway, is epistatic to *fz* (Krasnow and Adler 1994; Krasnow et al. 1995). Recently, it has been suggested that the Dsh protein uses different domains to act in the Wg or Fz pathway, with the DEP domain implicated in PP and the PDZ domain being required for Wg signaling (Axelrod et al. 1998; Boutros et al. 1998). This would suggest that part of the specificity of Fz vs Wg signaling in the cell is controlled by Dsh, which may interact differentially with two different receptors. More recent data showing that Fz itself can also act as a Wg receptor (see below) indicate that the situation is more complicated.

There are various other genes with PP phenotypes that either act in parallel to the Fz pathway, or are thought to be involved in signal transduction downstream of Fz (Fig. 3). In addition to *dsh*, the small GTPase RhoA as well as JNK have been implicated in signal transduction. *strabismus* (*stbm*)/*Van Gogh* may act in parallel, as well as *prickle* (*pk*), *dachsous* (*ds*), and *inturned* (*in*).

1.7 Interactions Between Wg, other Wg Signaling Components, and Dfz2 in Development

We obtained evidence that Dfz2 can act as a receptor for Wg during development by a dominant-negative approach, i.e., by expressing the extracellular domain of the Dfz2 protein as a GPI-linked cell surface protein. The wings of the resulting flies have marked defects in the margin, which is known to be specified by Wg, and in other structures such as the eyes and the legs. All the phenotypes observed are similar to loss of Wg function, which suggests that the extracellular domain binds Wg and inhibits its function.

Recently, several papers have taken genetic approaches to examine the loss-of-function phenotype of *Dfz2* (Bhat 1998; Kennerdell and Carthew 1998; Muller et al. 1999). It was found that *Dfz2* by itself is not required in the *Drosophila* embryo, but that double mutants between *fz* and *Dfz2* have the same phenotype as *wg*, showing that these two genes are genetically redundant.

1.8 Misexpression of *Dfz2*

To further examine the role of *Dfz2* in vivo, we studied its expression pattern in the developing wing and the consequences of misexpression of the intact receptor. In the wing pouch, the region of the disc destined to become wing blade, Dfz2 is expressed in an inverse concentration gradient to that of Wg, with the lowest levels found at the D/V boundary (Cadigan et al. 1998). This pattern is Wg-dependent, since Dfz2 expression is elevated when Wg activity is inactivated. Conversely, expression of an activated form of Arm throughout the wing pouch represses Dfz2 expression. Thus, Wg signaling is responsible for the graded expression of Dfz2.

To test whether the lowering of Dfz2 expression is important for normal wing development, we misexpressed Dfz2 using UAS-Dfz2 lines crossed to various Gal4 drivers (Brand and Perrimon 1993). Surviving animals have ectopic bristles on their wing blades. These sensory organs are normally found only at the wing margin, the adult structure corresponding to the D/V boundary, and depend on Wg activity for their formation (Couso et al. 1994; Phillips and Whittle 1993). The ectopic bristles in the anterior compartment were almost always of the slender or chemosensory type. The slender and chemosensory bristle cell fates are determined during the 3rd larval instar by proneural genes such as *ac*, whose expression is Wg-dependent. *ac* is initially expressed at mid-3rd instar in the anterior compartment in a stripe on each side of the D/V boundary. Consistent with the hairy wing phenotype, *1J3/Dfz2* discs have a dramatic increase in cells expressing high levels of Ac (Cadigan et al. 1998). These cells are found at a greater distance from the D/V stripe than in controls and presumably cause the ectopic bristles seen in adult wings.

Thus, misexpression of Dfz2 at high levels throughout the wing pouch expands the domains of both short- and long-range Wg targets.

1.9 The Dfz2 Phenotype is Wg-dependent

Increased activation of Wg targets by misexpression of Dfz2 could be due to an increased response of the cells to the Wg signal, or a constitutive activation of the signaling pathway. To address this, we examined the effect of Dfz2 misexpression in discs from wg^{ts} mutants reared at the restrictive temperature. Both Ac and Dll expression was dramatically reduced in these discs to levels seen in wg^{ts} discs grown under the same conditions.

These findings show that receptor phenotypes in vivo are dependent on the ligand and they further support the evidence that Dfz2 and Wg interact in vivo.

Acknowledgements. These studies were supported by the Howard Hughes Medical Institute.

References

Adler P N (1992) The Genetic Control of Tissue Polarity in *Drosophila*. Bioessays 14:735—741

Alcedo J, Ayzenzon M, Vonohlen T, Noll M, Hooper JE (1996) The *Drosophila smoothened* gene encodes a seven-pass membrane protein, a putative receptor for the Hedgehog signal. Cell 86:221—232

Axelrod JD, Miller JR, Shulman JM, Moon RT, Perrimon N (1998) Differential recruitment of Disheveled provides signaling specificity in the planar cell polarity and Wingless signaling pathways. Genes Dev 12:2610—2622

Behrens J, Von Kries JP, Kuhl M, Bruhn L, Wedlich D, Grosschedl R, Birchmeier W (1996) Functional interaction of β-catenin with the transcription factor LEF-1. Nature 382:638—642

Bhanot P, Brink M, Harryman Samos C, Hsieh J C, Wang YS, Macke JP, Andrew D, Nathans J, Nusse R (1996) A new member of the frizzled family from *Drosophila* functions as a Wingless receptor. Nature 382:225—230

Bhat KM (1998) frizzled and frizzled 2 play a partially redundant role in wingless signaling and have similar requirements to wingless in neurogenesis [In Process Citation]. Cell 95:1027—36

Boutros M, Paricio N, Strutt DI, Mlodzik M (1998) Disheveled activates JNK and discriminates between JNK pathways in planar polarity and wingless signaling. Cell 94:109—118

Brand AH, Perrimon N (1993) Targeted gene expression as a means of altering cell fates and generating dominant phenotypes. Development 118:401—415

Cadigan KM, Nusse R (1997) Wnt signaling: a common theme in animal development. Genes Dev 11:3286—3305

Cadigan KM, Fish MP, Rulifson EJ, Nusse R (1998) Wingless repression of *Drosophila* frizzled 2 expression shapes the Wingless morphogen gradient in the wing. Cell 93:767—777

Couso JP, Bishop SA, Martinez Arias A (1994) The wingless signalling pathway and the patterning of the wing margin in *Drosophila*. Development 120:621—636

Dohlman HG, Thorner J, Caron MG, Lefkowitz RJ (1991) Model systems for the study of seven-transmembrane-segment receptors. Annu Rev Biochem 60:653—88

Gomperts SN (1996) Clustering membrane proteins: it's all coming together with the PSD-95/SAP90 protein family. Cell 84:659—662

Gubb D, Garcia Bellido A (1982) A genetic analysis of the determination of cuticular polarity during development in *Drosophila melanogaster*. J Embryol Exp Morphol 68:37—57

Kennerdell JR, Carthew RW (1998) Use of dsRNA-mediated genetic interference to demonstrate that frizzled and frizzled 2 act in the wingless pathway [In Process Citation]. Cell 95:1017—26

Krasnow RE, Adler PN (1994) A single *frizzled* protein has a dual function in tissue polarity. Development 120:1883—1893

Krasnow RE, Wong LL, Adler PN (1995) Dishevelled is a component of the frizzled signaling pathway in *Drosophila*. Development 121:4095—4102

Miller JR, Moon RT (1996) Signal transduction through β-catenin and specification of cell fate during embryogenesis. Genes Dev 10:2527—2539

Molenaar M, Van de Wetering M, Oosterwegel M, Petersonmaduro J, Godsave S, Korinek V, Roose J, Destree O, Clevers H (1996) XTcf-3 transcription factor mediates β-catenin-induced axis formation in *Xenopus* embryos. Cell 86:391—399

Muller H, Samanta R, Wieschaus E (1999) Wingless signaling in the *Drosophila* embryo: zygotic requirements and the role of the frizzled genes. Development 126:577—586

Nusse R (1997) A versatile transcriptional effector of wingless signaling. Cell 89:321—323

Peifer M (1997) Enhanced β-catenin as oncogene— the smoking gun. Science 275:1752

Phillips RG, Whittle JRS (1993) *wingless* expression mediates determination of peripheral nervous system elements in late stages of *Drosophila* wing disc development. Development 118:427—438

Probst WC, Snyder LA, Schuster DI, Brosius J, Sealfon SC (1992) Sequence alignment of the G-protein coupled receptor superfamily. DNA Cell Biol 11:1—20

Shulman JM, Perrimon N, Axelrod JD (1998) Frizzled signaling and the developmental control of cell polarity. Trends Genet 14:452—8

Slusarski DC, Corces VG, Moon RT (1997) Interaction of Wnt and a frizzled homologue triggers G-protein-linked phosphatidylinositol signalling. Nature 390:410—413

Van Den Heuvel M, Ingham PW (1996) *smoothened* encodes a receptor-like serpentine protein required for *hedgehog* signalling. Nature 382:547—551

Van Leeuwen F, Harryman Samos C, Nusse R (1994) Biological activity of soluble wingless protein in cultured *Drosophila* imaginal disc cells. Nature 368:342—344

Vinson CR, Adler PN (1987) Directional non-cell autonomy and the transmission of polarity information by the frizzled gene of *Drosophila*. Nature 329:549—51

Wang Y, Macke J, Abella B, Andreasson K, Worley P, Gilbert D, Copeland N, Jenkins N, Nathans J (1996) A large family of putative transmembrane receptors homologous to the product of the *Drosophila* tissue polarity gene frizzled. J Biol Chem 271:4468—4476

Wodarz A, Nusse R (1998) Mechanisms of Wnt signaling in development. Ann Rev Cell Dev Biol 14:59—88

Yanagawa S, Van Leeuwen F, Wodarz A, Klingensmith J, Nusse R (1995) The Dishevelled protein is modified by Wingless signaling in *Drosophila*. Genes Dev 9:1087—1097

2 The Heterotrimeric G Protein Genes of Caenorhabditis elegans

G. Jansen, K. L. Thijssen, P. Werner, M. van der Horst,
E. Hazendonk, and R. H. A. Plasterk

2.1 Introduction

Caenorhabditis elegans is the first animal, and the first multicellular organism, for which the complete genomic sequence has been determined (The *C. elegans* Sequencing Consortium 1998). One of the new possibilities in post-sequence genetics is the immediate analysis of complete gene families. A first approximation analysis of gene function involves the determination of expression patterns, and the description of loss-of-function as well as gain-of-function phenotypes. We performed such studies for the complete family of G protein α subunits.

Heterotrimeric guanine nucleotide-binding proteins (G proteins) are at the interface between incoming extracellular signals received by a variety of cell surface receptors and intracellular effectors (reviewed in Neer 1995). As such, they play a pivotal role in the interaction of a cell with its environment. Due to the complexity of G protein-coupled signal transduction, it is still poorly understood how specificity in different pathways is regulated and maintained and how information from different pathways can be integrated. Knowledge of the function of all G protein subunits of an organism can greatly increase our understanding of these processes.

Several $G\alpha$ subunit genes have been isolated from *C. elegans* by PCR, using degenerate oligonucleotide primers (Fino Silva and Plasterk 1990; Lochrie et al. 1991; Mendel et al. 1995; Brundage et al. 1996; Park et al. 1997), or in forward genetic screens (Ségalat et al. 1995; Roayaie et al. 1998). Among these genes are homologues of three of the four mammalian classes; $G_i/G_o\alpha$: *goa-1* (Mendel et al. 1995; Ségalat et al. 1995) and $G_s\alpha$: *gsa-1* (Park et al. 1997). Functional analysis showed that *goa-1* and *gsa-1* are expressed in many neurons and muscle cells, and that mutation of these genes and *egl-30* affects muscle activity (in egg laying as well as locomotion) (Mendel et al. 1995; Ségalat et al. 1995; Brundage et al. 1996; Korswagen et al. 1997).

In addition, four $G\alpha$ genes have been identified that were not clear members of the known mammalian G protein classes (Lochrie et al. 1991; Roayaie et al. 1998). Three of these new $G\alpha$ genes, *gpa-2, gpa-3,* and *odr-3*, were found to be expressed in a subset of the amphid neurons, 12 pairs of sensory neurons in the head (Zwaal et al. 1997; Roayaie et al. 1998). Further analysis of the three new $G\alpha$ genes indeed showed that they are involved in several different perception processes: taste, olfaction, nociception, and dauer formation (Zwaal et al. 1997; Roayaie et al. 1998).

Although most of the specificity of G protein function lies in the α subunit, the $\beta\gamma$ complex also has significant signaling capabilities. Until recently only one β subunit gene was identified in *C. elegans, gpb-1* (van der Voorn et al. 1990). Inactivation of this gene leads to embryonic lethality as a result of inappropriate orientation of the planes of cell division (Zwaal et al. 1996). We recently found a second β subunit gene, *gpb-2*, which does not seem to be required for viability (F Simmer, R Korswagen, G Jansen, K Thijssen, and RHA Plasterk, unpublished results). Thus far, two γ subunit genes have been identified, *gpc-1* and *2* (G Jansen, K Thijssen, and RHA Plasterk, unpublished results). It should be noted that the γ subunit is small and not well conserved, so that identification by sequence comparison is not trivial. Therefore, it is likely that additional γ subunit genes remain to be found.

The availability of the genomic sequence and the conservation of all α subunits during evolution enabled us to identify all $G\alpha$ genes. Remarkably, *C. elegans* has 20 α subunits, of which 16 cannot be classified in one of the mammalian classes. In this study we focus on the complete family of α subunit genes of the nematode.

Functional genomics should ideally follow closely in the tracks of structural genomics, in that the same advantages of scale are sought for gene function analysis as were found for DNA sequencing. In the case of phenotypic analysis this is probably impossible. Even the description of expression patterns is not so easily scaled up. Ideally one would need to detect expressed proteins, but the generation of large series of well-identified specific antibodies is a major undertaking, and in gene families there is ample chance of overlapping specificities. Therefore we chose to determine a first approximation of the expression patterns by gene fusion studies, realizing that these may not always identify the definitive pattern. This approach, however, can be scaled up: PCR-amplified putative promoter regions of the G protein genes were fused in frame to the GFP reporter gene (Chalfie et al. 1994; A Fire, J Ahnn, G Seydoux, and S Xu, personal communication). These fusion constructs were used to generate transgenic animals. This procedure allowed us to determine in a short period of time the expression patterns for all Gα genes.

Similarly, reverse genetic analysis by gene inactivation needed to be scaled up. For this purpose we developed a method for target-selected gene deletion after chemical mutagenesis (Jansen et al. 1997). The principle of this method is as follows: animals are mutagenized by a chemical that is known to generate deletions, and a collection of frozen mutant lines is established. A corresponding set of (pooled) DNA samples is available for inspection by PCR. This inspection uses a selective PCR with primer pairs that flank the gene of interest; deletions are observed as PCR products that are more easily amplified than the larger wild-type DNA. In some cases we also used a related method that employs the Tc1 transposon as mutagen, which was developed by us previously (Zwaal et al. 1993).

Finally, we analyzed the effect of gain-of-function mutations for most genes by overproduction of the wild-type proteins in transgenic animals (Mello et al. 1991). To minimize aspecific effects we always used the endogenous promoter sequences to drive the expression of the overexpression or dominant active constructs. Since G proteins are molecular switches, one might expect that loss- and gain-of-function mutations have similar effects on signaling activity. It was previously described for *odr-3*, also a Gα subunit, that overproduction resulted in the same defect of odor perception as gene loss did (Roayaie et al. 1998). Alternatively, loss- and gain-of-function mutations might have opposite effects, as was found for *gpa-2* and *-3* (Zwaal et al. 1997). In cases where loss-of-function mutations may exhibit little phenotype due to redundancy, gain-of-function mutants may sometimes be more informative.

Taken together, these studies lead us to conclude that three quarters of the G protein α subunits of *C. elegans* act in perception: they are expressed in chemosensory neurons and mutations affect chemoattraction or aversion behavior (Jansen et al. 1999).

2.2 Results and Discussion

2.2.1 The *C. elegans* Genome Contains 20 Gα Genes

The *C. elegans* genomic sequence was screened using previously identified Gα gene sequences as query. This identified 20 G protein α subunit genes, named *goa-1, gsa-1, egl-30, gpa-1* to *-16*, and *odr-3*. Most Gα genes can be found on chromosome I or V (Fig. 1). There is a cluster of six Gα genes on chromosome V, spread over 4.5 map units. Four of them are quite close together, within 0.5 map units, of which *gpa-8* and *-9* are only approximately 6 kb apart. No Gα genes have been identified on chromosome III.

The predicted amino acid sequences of all *C. elegans* Gα genes were aligned and compared to representatives of the four mammalian classes: $G_i/G_o\alpha$, $G_s\alpha$, $G_q\alpha$, and $G_{12}\alpha$ (Fig. 2). The overall structure, a GTPase domain joined through two linker peptides to an α-helical domain, seems to be conserved between these Gα subunits. There are however substitutions of presumed key amino acid residues in *gpa-5, -11, -13,* and *-14*. How these alterations affect G-protein function is unknown.

The alignment showed that there is only one homologue for each of the four main vertebrate classes of Gα genes (Fig. 2); $G_i/G_o\alpha$: *goa-1* (Lochrie et al. 1991), $G_s\alpha$: *gsa-1* (Park et al. 1997), $G_q\alpha$: *egl-30* (Brundage et al. 1996), and $G_{12}\alpha$: *gpa-12*. As the 16 other Gα subunits cannot

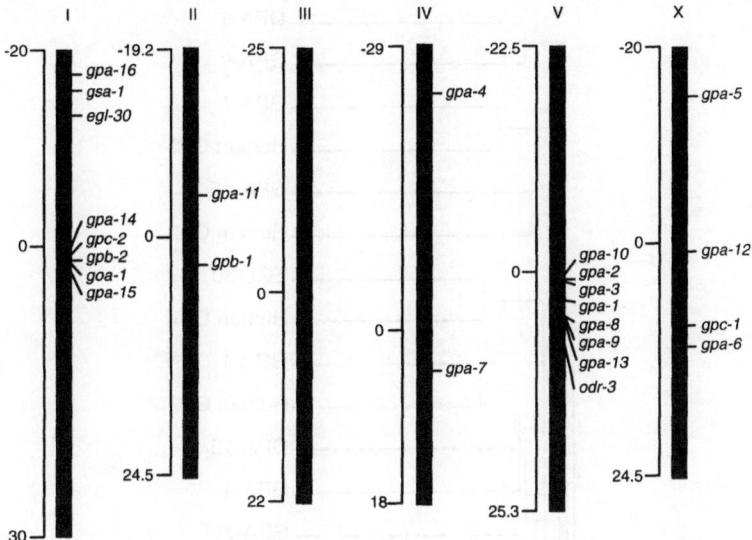

Fig. 1. Genomic localization of all Gα, Gβ and Gγ genes, extrapolated to the genetic map of *C. elegans*. There is some clustering of Gα genes on chromosome *I* and *V*. The six chromosomes are presented by *black bars*, *left* of each chromosome is its size (given in map units). The approximate position of all G protein genes is indicated to the *right* of each chromosome

clearly be grouped in any of these classes, or a new class, we will consider these 16 genes as an outgroup of new Gα genes.

2.2.2 Most Gα Genes Are Expressed in Chemosensory Cells

As a first step in the functional characterization of all Gα subunits of *C. elegans* the expression pattern for each of these genes was determined. The expression patterns of *goa-1, gsa-1, gpa-2, gpa-3*, and *odr-3* have been previously reported (Table 1 and references therein). Transgenic animals were generated that express the reporter gene *GFP* (green fluorescent protein; Chalfie et al. 1994; A. Fire, J. Ahnn, G. Seydoux, and S. Xu, personal communication) driven by the putative promoter of each of the 13 Gα genes tested (i.e., all but *egl-30* and *gpa-12*).

G. Jansen et al.

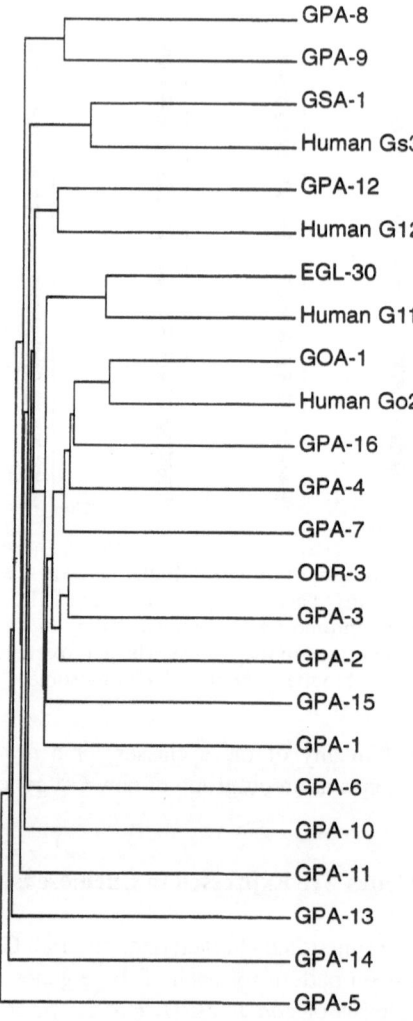

Fig. 2. Alignment dendrogram generated using the PILEUP algorithm using default program parameters. The predicted amino acid sequences of all $G\alpha$ subunits of *C. elegans* were compared to human $G_{s3}\alpha$, $G_{12}\alpha$, $G_{11}\alpha$, and $G_{o2}\alpha$

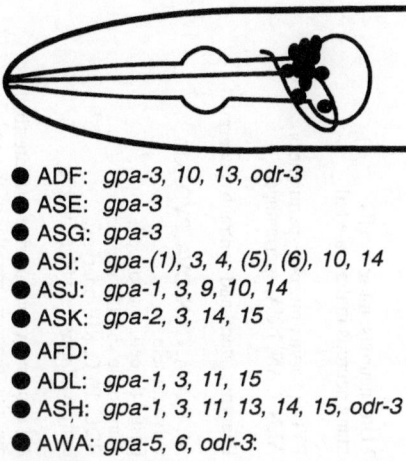

● ADF: *gpa-3, 10, 13, odr-3*
● ASE: *gpa-3*
● ASG: *gpa-3*
● ASI: *gpa-(1), 3, 4, (5), (6), 10, 14*
● ASJ: *gpa-1, 3, 9, 10, 14*
● ASK: *gpa-2, 3, 14, 15*
● AFD:
● ADL: *gpa-1, 3, 11, 15*
● ASH: *gpa-1, 3, 11, 13, 14, 15, odr-3*
● AWA: *gpa-5, 6, odr-3*:
● AWB: *odr-3*
● AWC: *gpa-2, 13, odr-3*

Fig. 3. Summary of the amphid neuron specific expression in a schematic drawing of the head of *C. elegans*. The cell bodies of all amphid neurons are indicated. The processes to the amphid sensilla in the tip of the nose and into the nerve ring of one amphid neuron are drawn. The expression of the Gα subunits in each of the cells as determined in this study, or described for *gpa-2, -3* (Zwaal et al. 1997), and *odr-3* (Roayaie et al. 1998), is given

Of the 20 Gα fusions, 13 are expressed in a small subset of sensory neurons, the amphid cells (Fig. 3; Table 1). These sensory neurons in the head are involved in perception of volatile attractants (AWA, AWC) and repellents (ASH, AWB, ADL), water soluble attractants (ASE, ADF, ASG, ASI, ASK) and repellents (ASH, ADL), the dauer pheromone (ADF, ASG, ASI, ASJ), nose touch (ASH), and temperature (AFD) (reviewed in Bargmann and Mori 1997). In addition to the amphid expression several genes showed expression in other cells (Table 1), frequently in the phasmid neurons, putative sensory neurons (White et al. 1986).

In contrast, *gpa-7::GFP* staining was observed in most neurons in the head and tail ganglia, and in the body. Furthermore, weak *gpa-7::GFP* expression was seen in all muscle cells. *gpa-16::GFP* staining was observed in several neurons, including the AVM and PLM touch neurons, faintly in various muscle cells, and in the pharynx.

Table 1. Summary of GFP expression data

Gene	Ref.	Amphid cells	Other putative sensory cells	Other expression
gpa-1		ADL, ASH, ASI (faint), ASJ	PHA, PHB	1 or 2 pairs of pcs neurons in the male tail
gpa-2	1	AWC	PHA, PHB, and IL1L, IL2L, OLL, or URB	PVT, anal sphincter muscle, M1, M5, 15 neurons, AIA
gpa-3	1	ADF, ADL, ASE, ASG, ASH, ASI, ASJ, ASK	PHA, PHB	PVT, AIZ
gpa-4		ASI		
gpa-5		AWA, ASI (faint)		
gpa-6		AWA, ASI (faint)	PHB	
gpa-7				Many neurons, muscle cells, many neurons in the male tail
gpa-8			URX, AQR, PQR	
gpa-9		ASJ	PHB	PVQ, pharynx muscle, spermatheca
gpa-10		ADF, ASI, ASJ		ALN, CAN, LUA, spermatheca
gpa-11		ADL, ASH		
gpa-12	2			Pharynx, lateral and ventral hypodermis
gpa-13		ADF, ASH, AWC	PHA, PHB	
gpa-14		ASI, ASJ, ASH, ASK	ADE, PHA, PHB	ALA, AVA, CAN, DVA, PVQ, RIA, vulva muscle
gpa-15		ADL, ASH, ASK	PHA, PHB	Distal tip cell, anchor cell, many male specific neurons
gpa-16			AVM, PDE, PLM	BDU, PVC, RIP, pharynx, body wall muscle, vulva muscle
odr-3	3	AWA, AWB, AWC, ASH, ADF		
goa-1	4, 5	All	All	All neurons, vulva, uterine and intestinal muscles, diagonal muscles of the male, most cells of the pharynx, distal tip cells
gsa-1	6, 7	All	All	All neurons, most muscle cells, excretory cell
egl-30		Unknown	Unknown	Unknown

The cellular pattern of expression of each gene is given, as observed with gene specific GFP fusion constructs. The expression pattern of several Gα genes was reported previously: 1, Zwaal et al. 1997; 2, L. Brundage and M. Simon personal communication; 3, Roayaie et al. 1998; 4, Mendel et al. 1995; 5, Segalat et al. 1995; 6, Park et al. 1997; 7, Korswagen et al. 1997. pcs, Postcloacal sensillae.

The *GFP* fusion constructs of *gpa-1*, *gpa-7*, *gpa-13*, and *gpa-15* showed male-specific expression (Table 1), in addition to the expression seen in hermaphrodites; the others showed identical staining patterns in males and hermaphrodites. The expression patterns suggest that *C. elegans* uses at least 14 new Gα genes in chemosensory neurons.

2.2.3 The New Gα Genes Are Not Essential

To further analyze the function of all Gα genes in *C. elegans* we isolated loss-of-function alleles (Fig. 4), using target selected gene inactivation (Zwaal et al. 1993; Jansen et al. 1997). Gain-of-function alleles were generated by introducing the wild-type gene at elevated gene copy number *(XS)* as transgenes. All mutants generated in this study and the previously isolated mutants *gpa-2*, *gpa-2QL*, *gpa-3*, *gpa-3QL* (Zwaal et al. 1997), and *odr-3* (Roayaie et al. 1998) were tested for phenotypes (Table 2). The QL mutants overexpress a constitutively active mutant gene that codes for a protein with strongly reduced GTPase activity. None of the mutations tested in this study resulted in a lethal phenotype. Based on the expression patterns, all mutants were tested for subtle phenotypes.

The *gpa-7::GFP* expression suggests a function of *gpa-7* in regulating muscle or neuron activity, as described for *goa-1*, *gsa-1*, and *egl-30* (Mendel et al., 1995; Ségalat et al. 1995; Brundage et al. 1996; Korswagen et al. 1997). *gpa-7* animals were egg-laying defective and *gpa-7XS* animals were hyperactive regarding egg laying, as measured by the number of eggs in utero (wild type 12.0±0.6; *gpa-7* 23.4±1.2; *gpa-7XS* 6.7±0.6; number of eggs in utero±SEM, in each case *n*=25). No significant difference could be observed in body wall muscle activity. These results indicate that *gpa-7* possesses a stimulatory function in muscle cells and/or neurons. Similar functions have been ascribed to *gsa-1* (Korswagen et al. 1997) and *egl-30* (Brundage et al. 1996). It is interesting to see that although GPA-7 is most similar to G$_o$α, mutations in *gpa-7* result in phenotypes opposite to *goa-1*, which inhibits the activity of muscles and neurons (Mendel et al. 1995; Ségalat et al. 1995). Taken together, these data indicate that at least four Gα subunits regulate muscle and neuron activity in *C. elegans*: three of the four mammalian homologues and *gpa-7*.

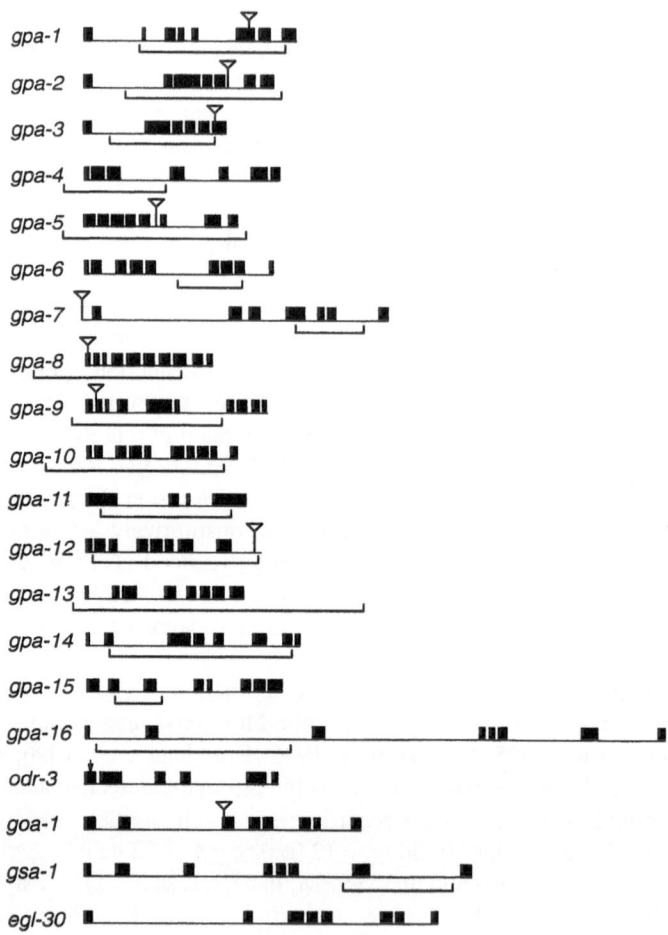

Fig. 4. Predicted gene structure and mutant alleles of the G protein α subunit genes of *C. elegans*. Exons are represented by *black boxes*. For some genes transposon Tc1 insertion alleles have been isolated, indicated with *triangles*. *Brackets* show the regions deleted in the knockout alleles. The gene structure and mutant alleles of several Gα genes have been reported previously: *gpa-2* and *-3* (Zwaal et al. 1997); *odr-3* (Roayaie et al. 1998), the position of the non-sense mutation present in allele *n1605* is indicated with an *arrow*; *goa-1* (Lochrie et al. 1991), the Tc1 insertion in allele *pk62* is indicated (Mendel et al. 1995); *gsa-1* (Park et al. 1997), the deletion present in allele *pk75* is given (Korswagen et al. 1997); *egl-30* (Brundage et al. 1996), many different alleles exist, none of these are indicated

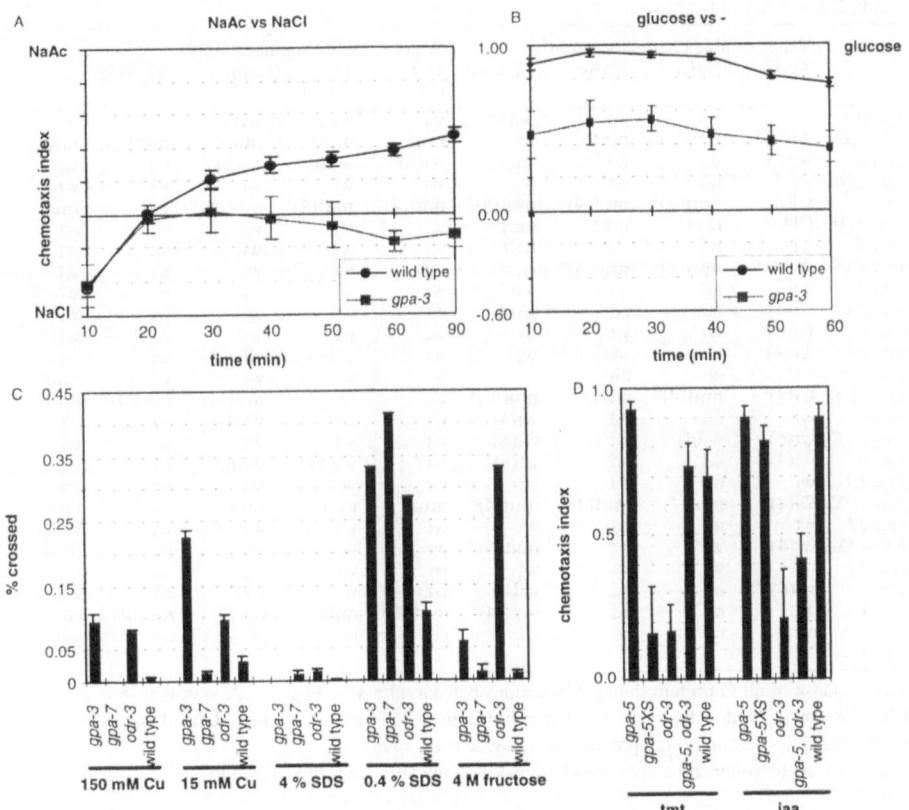

Fig. 5A–D. Six Gα knockout alleles show chemotaxis defects. **A,B,** *gpa-3* seems involved in chemotaxis to water soluble compounds. Mutant or wild-type animals are given a choice between NaAc or NaCl (**A**), or are tested for chemotaxis to glucose (**B**). The chemotaxis index in time is plotted. A chemotaxis index of 1 means complete attraction to NaAc (**A**) or glucose (**B**), whereas –1 means complete attraction to NaCl (**A**) or avoidance of glucose (**B**). **C** *gpa-3, -7,* and *odr-3* seem involved in water-soluble chemoaversion. Animals were tested for avoidance of 150 mM and 15 mM $CuSO_4$, 4% and 0.4% SDS, and 4 M fructose. The percentages of animals that crossed the aversive barrier, attracted by isoamylalcohol, are plotted for each mutant. **D** *gpa-5* is a negative regulator of 2,4,5-trimethylthiazole (*tmt*) perception. The chemotaxis index is plotted for 100 times diluted *tmt* and isoamyl alcohol (*iaa*)

Table 2. Summary of behavioral assays

Strain	Water soluble chemoattraction[a]				Water soluble chemoaversion[b]				
	Ac-Cl	NaCl	NaAc	Glucose	150 CuSO$_4$	15 CuSO$_4$	osm	4% SDS	
gpa-1	wt	nd	nd	wt(4)	wt	wt	wt	wt	wt
gpa-1XS	**Ac(4)***	**mut***	**mut***	wt	**mut***	**mut***	**mut***	**mut***	**mut***
gpa-2	wt(4)	nd	nd	wt(4)	wt(6)	wt(6)	wt(5)	wt	wt
gpa-2QL	wt	nd	nd	wt	wt	wt	wt	wt	wt
gpa-3	**Cl(4)***	**mut(4)***	**mut(4)***	**mut(8)***	**mut(4)***	**mut(4)***	**mut(8)***	wt	**mut***
gpa-3QL	**Cl***	**mut***	**mut**	**mut***	wt	wt	wt	nd	nd
gpa-4	wt	nd	nd	wt(6)	wt	wt	wt	wt	wt
gpa-4XS	**Ac(4)***	**mut(4)***	**mut(4)***	wt(4)	wt	wt	wt	wt	wt
gpa-5	wt	nd	nd	wt(4)	wt	wt	wt	wt	wt(4)
gpa-5XS	wt(4)	nd	nd	wt(4)	wt	wt	wt	wt	wt
gpa-6	**Cl(6)***	wt	wt	wt	wt	wt	wt	wt	wt
gpa-6XS	**Ac(6)**	wt	wt	wt	wt	wt(4)	wt	wt	wt
gpa-7	wt(4)	nd	nd	wt	wt	wt	wt	wt	**mut***
gpa-7XS	**Cl(4)***	**mut(4)***	wt(4)	**mut(4)***	wt	wt	**mut(4)***	**mut(4)***	**mut(4)***
gpa-8	wt	nd	nd	wt(4)	wt	wt	wt(4)	wt	wt
gpa-8XS	wt(6)	wt(4)	wt	wt(4)	wt	wt	wt	wt	wt
gpa-9	wt(4)	nd	nd	wt(8)	wt	wt(4)	wt(6)	wt	wt
gpa-10	wt	nd	nd	wt	wt	wt	wt	wt	wt
gpa-10XS	**Cl(4)***	**mut(4)***	**mut(4)***	**mut(4)***	**mut***	**mut**	**mut***	**mut***	**mut***
gpa-11	wt	nd	nd	wt	wt	wt	wt	wt	wt(4)
gpa-11XS	wt(4)	wt	wt	**mut(6)***	wt	wt	wt	wt	wt(4)
gpa-14	wt	wt	wt	wt	wt	wt	wt	wt	wt
gpa-15	wt(6)	nd	nd	wt(4)	wt	wt	wt	wt	wt(4)
odr-3	wt	nd	nd	**mut(4)***	**mut(4)***	**mut***	**mut***	**mut(4)**	nd
gpa-5, odr-3	nd	nd	nd	nd	nd	nd	nd	nd	nd

The response of all G protein mutants in various behavioral assays is given. Assays that showed altered behavior are in bold. All assays were performed twice, unless otherwise indicated (*n*). Some *odr-3* phenotypes have been reported previously (Roayaie et al. 1998).

wt, wild type behavior; mut, reduced chemotaxis to the attractant, or reduced avoidance; nd, not determined.

*Mutants that showed significantly altered behavior (*P*.05).

[a] Chemotaxis to water soluble compounds. Ac-Cl, animals were presented with a choice between NaAc and NaCl. Increased preference for Ac (Ac), or Cl (Cl) is indicated. Chemotaxis to NaCl, NaAc, and glucose.

[b] Aversion of water soluble compounds, 150 mM or 15 mM CuSO$_4$; 4 M fructose (osm); 4% SDS, 0.4% SDS.

[c] Chemotaxis to volatile attractants, trimethylthiazole (tmt), diacetyl (dia), pyrazine (pyr) (10 mg/ml), isoamyl alcohol (iaa), benzaldehyde (ben), pentanedione (pen). All odorants were diluted 100 times in ethanol, unless otherwise indicated.

[d] Avoidance of volatile compounds, non, 2-nonanone; oct, 1-octanol, undiluted. Assays performed using 10 times diluted odorants confirmed the results.

[e] *gpa-1XS* and *gpa-11XS* animals showed reduced potency (55% and 40% potency, respectively), 80–100% of the males of all other strains were potent.

[f] Animals were incubated with the fluorescent dye DiO, filling of amphid and phasmid neurons was tested. *gpa-10XS* animals showed stronger staining, *gpa-3QL* animals showed no filling.

Table 2. Continued

Volatile chemoattraction[c]						Volatile aversion[d]		Male potency[e]	DiO filling[f]
tmt	dia	pyr	iaa	ben	pen	non	oct		
wt	wt(4)	wt	wt(4)	wt	wt	wt	wt	wt	wt
wt	wt	wt	wt	wt	wt	wt(4)	wt	**reduced**	wt
wt	wt	wt(4)	wt(4)	wt	wt	wt	wt	wt	wt
wt	wt	**mut(4)**	**mut(4)**	**mut(4)***	wt	**mut***	**mut***	wt	wt
wt(4)	wt	**mut(6)**	wt	wt	wt	wt	wt	wt	wt
mut*	**mut***	**mut(2)**	**mut***	nd	nd	**mut***	nd	wt	no
wt(6)	wt(6)	wt(4)	wt	wt(4)	wt(4)	wt	wt(4)	wt	wt
wt	wt	wt	wt	wt	wt	wt	wt	wt	wt
wt(6)	wt(6)	wt(6)	wt(7)	wt(4)	wt	wt(4)	wt	wt	wt
mut(6)*	**mut(6)***	**mut(6)***	wt(7)	wt	wt	wt	wt(4)	wt	wt
wt	wt	wt	wt	wt	wt	wt	wt	wt	wt
wt	wt	wt(4)	wt	wt(4)	wt	wt	wt	wt	wt
wt	wt	wt	wt	wt	wt	wt	wt	wt	wt
mut(6)*	wt	**mut(4)**	**mut(4)***	wt	**mut(4)**	**mut(4)***	wt	wt	wt
wt(6)	wt(6)	wt(4)	wt	wt	wt	wt	wt(6)	wt	wt
wt(4)	wt	wt	wt	wt	wt	wt	wt	wt	wt
wt	wt	wt	wt	wt	wt	wt(4)	wt	wt	wt
wt	wt(4)	wt	wt	wt	wt	wt	wt(4)	wt	wt
wt(4)	wt(4)	**mut(4)**	wt(4)	wt(4)	wt(4)	**mut***	wt(4)	wt	**stronger**
wt	wt	wt	wt	wt(4)	wt	wt	wt(6)	wt	wt
wt	wt(4)	wt	wt	wt	wt	wt	wt	reduced	wt
wt	wt	wt	wt(6)	wt(8)	wt(6)	wt	wt(6)	wt	wt
wt	wt	wt	wt(4)	wt	wt	wt	wt	wt	wt
mut(6)*	**mut(6)***	**mut(6)***	**mut(7)***	**mut**	nd	**mut**	**mut**	nd	wt
wt(6)	**mut(6)***	**mut(6)***	**mut(7)***	nd	nd	nd	nd	nd	nd

2.2.4 Several Gα Genes Are Involved in Perception

The expression of 14 of the 16 new Gα genes in amphid neurons or other putative sensory neurons, suggests they might be involved in perception, or in the development of sensory neurons, as has been reported for ODR-3 (Roayaie et al. 1998) and GPA-3 (Zwaal et al. 1997).

We tested whether the endings of the amphid and phasmid neurons are still exposed to the environment by dye filling (Perkins et al.1986). None of the mutants tested, except *gpa-3QL* animals (Zwaal et al. 1997), showed reduced dye filling (data not shown). Surprisingly, *gpa-10XS* animals showed stronger dye filling (data not shown). This phenotype may be the result of structural changes in the sheath cells.

The ability of the mutants to sense and respond to several water soluble or volatile stimuli was tested: chemotaxis to sodium acetate (NaAc), NaCl and glucose (Ward 1973), avoidance of coppersulfate, sodium dodecylsulfate (SDS) and 4 M fructose (osmotic avoidance; Culotti and Russell 1978; Bargmann and Mori 1997), and attraction to, or aversion of, several different volatile compounds (Bargmann et al. 1993; Sengupta et al. 1994; Troemel et al. 1997). Furthermore, male potency was tested (Mendel et al. 1995). Surprisingly, we found only few phenotypes for loss-of-function alleles. Three genes seem involved in chemotaxis to water soluble compounds: *gpa-3*, *gpa-6*, and *odr-3* (Table 2; Fig. 5A,B). Three loss-of-function mutations affect avoidance behavior: *gpa-3*, *gpa-7*, and *odr-3* (Roayaie et al. 1998) (Table 2; Fig. 5C). Also, three genes seem to be involved in olfaction: *gpa-2* (Roayaie et al. 1998), *gpa-5*, and *odr-3* (Roayaie et al. 1998) (Table 2; Fig. 5D). None of the loss-of-function alleles showed reduced male potency.

In contrast, eight of the ten gain-of-function alleles tested showed phenotypes in these assays (Table 2). Although we cannot extrapolate gene function from dominant mutants with any certainty, these phenotypes can provide hints about the involvement of G proteins in specific cellular functions. They show that there are probably independent cells and pathways for acetate/Cl and glucose perception (Table 2). The ASI neurons would be likely candidates for the perception of NaAc and NaCl, the ADF and ASH neurons would be likely, but not perfect, candidates for glucose perception. Furthermore, there are probably independent pathways for the three repellents tested. The most likely candidate cells involved in aversion are ADL and ASH (Bargmann and Mori 1997). Finally, the effect of overexpression of GPA-5 on olfaction led us to test functional redundancy between *gpa-5* and *odr-3*. Surprisingly, a double mutant of *gpa-5* and *odr-3* showed wild-type chemotaxis to 2,4,5-trimethylthiazole (tmt), but reduced chemotaxis to other odorants (Table 2; Fig. 5D), suggesting that *gpa-5* is a negative regulator of tmt perception.

Mutations in only two Gα subunits, *gpa-3* and *odr-3*, have profound effects on perception. The other genes probably have more subtle functions, which could be masked by functional redundancy, as was described for *odr-3* and *gpa-2* (Roayaie et al. 1998). Given the expression patterns of multiple α subunit genes per neuron this seems certainly possible. Alternatively, these genes could be involved in behaviors that have not yet been tested, for example perception of other compounds, or they may have negative regulatory functions (like *gpa-5*). It will be interesting to see if a similar ratio of central to regulatory genes holds true also in other systems. Perhaps forward genetics has biased our focus in functional analysis to genes with clear phenotypes, and has caused an underestimation of the importance of regulatory pathways.

2.2.5 Why Would *C. elegans* Need 14 or More Gα Subunits for Perception?

The answer to this question probably resides in the simplicity of the animal. Mammals are thought to have each of their millions of olfactory neurons devoted to one specificity (Buck 1996). The adult *C. elegans* hermaphrodite has only 16 pairs of neurons that are in direct contact with the animal's environment. However, the animal senses much more than 16 compounds, and responds specifically to different stimuli perceived via the same cell (Bargmann et al. 1993; Kaplan and Horvitz 1993; Colbert and Bargmann 1995). To do so, *C. elegans* needs multiple signal transduction pathways per cell. Analyses of the expression patterns of some of the approximately 650 putative serpentine receptor genes (Troemel et al. 1995) and 29 guanylyl cyclases (Yu et al. 1997) indicate that indeed, each chemosensory cell expresses several receptor genes. Our data suggest that the nematode also uses multiple Gα subunit genes per cell. We can now address questions such as how specificity is achieved and maintained in the presence of so many, probably rather similar signal transduction cascades, and where the information is integrated.

2.3 Methods

2.3.1 Gα Gene Identification

The *C. elegans* genomic sequence, available from the Genome Sequencing Centers (The Sanger Centre, Cambridge, UK, and Washington University, St. Louis, Mo. USA), was repeatedly screened for Gα subunit genes in a TBLASTN search (Altschul et al. 1990).The DNA sequences were further analyzed using the Wisconsin Sequence Analysis Package. The predicted gene structures were further refined using the GENEFINDER predictions, as annotated in the *C. elegans* database ACeDB (Eeckman and Durbin 1995). Discrepancies between the GENEFINDER predictions and ours were communicated to those annotating the sequences.

We identified 20 G protein α subunit genes in the *C. elegans* genome (Jansen et al. 1999): *gpa-1* (Lochrie et al. 1991), located on cosmid T19C4, gene 6 (i.e., T19C4.6); *gpa-2* (Zwaal et al. 1997), F38E1.5; *gpa-3* (Zwaal et al. 1997), E02C12.5; *gpa-4*, T07A9.m; *gpa-5*, F53B1.7; *gpa-6*, F48C11.1; *gpa-7*, R10H10.5; *gpa-8*, F56H9.3; *gpa-9*, F56H9.4; gpa-10, C55H1.2; *gpa-11*, C16A11.1; *gpa-12*, F18G5.3; *gpa-13*, C15C8.7; *gpa-14*, B0207.3; *gpa-15*, M04C7.1; *gpa-16*, Y95B8; *odr-3* (Roayaie et al. 1998), C34D1.3; *goa-1* (Lochrie et al. 1991), C26C6.2; *gsa-1* (Park et al. 1997), R06A10.2; *egl-30* (Brundage et al. 1996), M01D7.7. Furthermore, two Gβ and two Gγ subunits were identified: *gpb-1* (Zwaal et al. 1996), F13D12.7; *gpb-2*, F52A8.2; *gpc-1*, K02A4.2; *gpc-2*, F08B6.e. None of these genes seems part of an operon.

The gene structure prediction of *gpa-5* and *-16* was confirmed by the sequencing of random and oligo-dT primed cDNA amplified by reverse transcriptase PCR.

2.3.2 GFP Fusion Constructs

Routinely, a 3 kbp PCR fragment, containing approximately 2.5–3 kbp of upstream sequences and the first 30–50 codons of the α genes, was fused in frame to the reporter gene *GFP* (vector pPD95.77; A. Fire, J. Ahnn, G. Seydoux, and S. Xu, personal communication). In most cases the first intron was present. In this way fusion constructs were generated for *gpa-4, -6, -7, -8, -9, -10, -11, -13, -14, -15* and, *-16*. Promoter fusions for *gpa-1* and *gpa-5* were generated by subcloning relevant regions from cosmid clones. The *gpa-1::GFP* fusion construct contains 1.5 kbp of upstream sequences, and the first eight exons of *gpa-1*. The *gpa-5::GFP* construct contains 3.7 kbp upstream sequences and the first five exons of *gpa-5* (Jansen et al. 1999).

At least two independent transgenic lines were generated from at least two independent clones of each of the *gpa::GFP* fusion constructs, to control for PCR induced sequence errors. Except for variation in expression level, all transgenic lines generated for each of the Gα subunits showed the same expression pattern. Subsequently, the transgenic array was integrated for one of the lines of each of the fusion constructs. Particular cells were identified by using a combination of their position and morphology (White et al. 1986). Expression of all *gpa::GFP* fusions was observed in animals from the embryonic three-fold stage onwards, or from the moment a postembryonic cell was born (data not shown). Expression in males was examined for all GFP fusion constructs except *gpa-16::GFP*.

2.3.3 Isolation of Loss-of-Function Alleles

Deletion mutations of *gpa-1, gpa-5,* and *gpa-8* were isolated as described: *gpa-1(pk15), gpa-5(pk376),* and *gpa-8(pk345)* (Zwaal et al. 1993). Deletion alleles of *gpa-4, -6, -7, -9, -10, -11, -14, -15,* and *-16* were isolated using a recently developed method (Jansen et al. 1997) that uses chemical mutagenesis: *gpa-4(pk381), gpa-6(pk480), gpa-7(pk610), gpa-9(pk438), gpa-10(pk362), gpa-11(pk349), gpa-12(pk322), gpa-13(pk482), gpa-14(pk347), gpa-15(pk477),* and *gpa-16(pk481)* (Jansen et al. 1999). Besides the deletion, the *gpa-13* allele seems to contain a complex rearrangement, which is currently under study. As all deletions were induced by random mutagenesis, all alleles were backcrossed six times to a wild type Bristol N2 background, before phenotypic analysis. The loss-of-function allele of *gpa-16* has only recently been isolated. As only a non-backcrossed allele was available, this mutant has not been characterized phenotypically.

2.3.4 Overexpression Constructs

Mutants that putatively overexpress one of the Gα genes (*gpa-1XS, 4XS, 5XS, 6XS, 7XS, 8XS, 10XS,* and *11XS)* were generated by introducing extra copies of the wild type gene as transgenes (Jansen et al. 1999). The Gα subunit genes were subcloned so that no other complete gene was present. Marker *dpy-20* DNA (Han and Sternberg 1991) was used at a concentration of 100 μg/ml and test DNA at a concentration of 50 μg/ml. The transgenic arrays were integrated by irradiating trans-genic animals with 40 Gy of γ radiation from a ^{137}Cs source. Before phenotypic analysis all transgenic strains were outcrossed at least twice.

2.3.5 Behavioral Assays

Nematode chemotaxis to soluble compounds was assessed by a newly developed assay (SRWicks and RHA Plasterk, in preparation). Briefly, young adult nematodes were placed on a plate at the intersection of four quadrants, filled with buffered agar either containing a dissolved attrac-tant (75 mM NaAc, 75 mM NaCl, or 75 mM glucose) or no attractant. The distribution of the worms over the four quadrants was determined at 10, 20, 30, 40, 50, 60, and 90 min. A chemotaxis index was calculated (CI=(A-C)/A+C: A is the number of worms over attractant X, C is the number of worms over attractant Y, or control). In this assay worms can be presented with the choice between an attractant and only buffered agar or with a choice between two salts (NaAc vs NaCl). In the latter case wild-type animals show a preference for Cl in the beginning, but in time prefer Ac (Fig. 5A). Subtle changes in the animals' behavior can now be detected more easily.

Soluble compound avoidance assays (C.J. de Vries and RHA Plas-terk, unpublished results) were performed using 150 mM and 15 mM

CuSO4, 4% and 0.4% SDS, and 4 M fructose as aversive compounds. Young adult worms were placed on a plate and stimulated to cross an aversive barrier by placing 1 μl of isoamylalcohol within a semicircle of the aversive compound. The animals on the two halves of the plate were counted after 90 min, and an aversion index was calculated as the percentage of animals that crawled through the barrier.

Volatile chemotaxis assays were performed as described (Bargmann and Horvitz 1991). Attractants used were: 2,4,5-trimethylthiazole, perceived by AWA and AWC; pyrazine (10 and 100 mg/ml) and diacetyl perceived by AWA; and isoamylalcohol, benzaldehyde and 2,3-pentanedione perceived by AWC (Bargmann et al. 1993; Sengupta et al. 1994). Aversive odorants used were 2-nonanone, detected by the AWB neurons (Troemel et al. 1997), and 1-octanol detected by ADL and ASH (Troemel et al. 1995). All attractants were tested in a 10 or 100 times dilution (unless otherwise indicated) in ethanol. The aversive odorants were tested undiluted or diluted ten times in ethanol.

Male mating efficiency was tested as described (Mendel et al. 1995). Locomotion and egg laying were assayed as described (Korswagen et al. 1997).

2.3.6 Statistical Analysis

Statistical analysis of behavioral data was done with SPSS 8.0 for Windows. Depending on the assay, a paired t-test, a one way ANOVA and Dunnett post-hoc, or factorial ANOVA and Tuckey post-hoc comparisons were used. An α level of 0.05 was used in all tests. All results are given as mean±SEM.

Acknowledgments. Part of the results presented in this paper have previously been reported in Jansen et al. 1999. We thank Cori Bargmann for strains and help in identification of amphid neurons, Lorna Brundage and Mel Simon for the communication of unpublished results, Alan Coulson for cosmids, and the Fire lab for GFP-expression vectors. We gratefully acknowledge Richard Zwaal, Jaap Neels, Rik Korswagen, and Yusuke Kato for the isolation of mutant strains, Corry de Vries for help with behavioral assays and Stephen Wicks for help with statistical analyses and behavioral assays. We thank Claudia van den Berg, Piet Borst, Rik Korswagen, and Stephen Wicks for comments on the manuscript. This work was supported by grant NKI 94–809 from the Netherlands Cancer Foundation, grant NWO-GMW 90104094 from the Netherlands Organization for Scientific Research, grant 940–70–008 from the New Drugs Research Foundation to RHAP, and by a Biotechnology Research Training Grant from the European Commission (BIO4CT965072) to PW.

References

Altschul SF, Gish W, Miller W, Myers EW, Lipman DJ (1990) Basic local alignment search tool. J Mol Biol 215:403–410

Bargmann CI, Horvitz HR (1991) Chemosensory neurons with overlapping functions direct chemotaxis to multiple chemicals in *C. elegans*. Neuron 7:729–742

Bargmann CI, Hartwieg E, Horvitz HR (1993) Odorant-selective genes and neurons mediate olfaction in *C. elegans*. Cell 74:515–527

Bargmann CI, Mori I (1997) Chemotaxis and thermotaxis. In: Riddle DL, Blumenthal T, Meyer BJ, Priess JR (eds) *C. elegans II*. Cold Spring Harbor Press, New York, pp 717–737

Brundage L, Avery L, Katz A, Kim U-J, Mendel JE, Sternberg PW, Simon MI (1996) Mutations in a *C. elegans* $G_q\alpha$ gene disrupt movement, egg-laying and viability. Neuron 16:999–1009

Buck LB (1996) Information coding in the vertebrate olfactory system. Annu Rev Neurosci 19:517–544

Chalfie M, Tu Y, Euskirchen G, Ward WW, Prasher DC (1994) Green fluorescent protein as a marker for gene expression. Science 263:802–805

Colbert HA, Bargmann CI (1995) Odorant-specific adaptation pathways generate olfactory plasticity in *C. elegans*. Neuron 14:803–812

Culotti JG, Russell RL (1978) Osmotic avoidance defective mutants of the nematode *Caenorhabditis elegans*. Genetics 90:243–256

Eeckman FH, Durbin R (1995) ACeDB and Macace. Methods Cell Biol 48:583–605

Fino Silva I, Plasterk RHA (1990) Characterization of a G protein α-subunit gene from the nematode *Caenorhabditis elegans*. J Mol Biol 215:483–487

Han M, Sternberg PW (1991) Analysis of dominant negative mutations of the *Caenorhabditis elegans let-60* ras gene. Genes Dev 5:2188–2198

Jansen G, Hazendonk E, Thijssen KL, Plasterk RHA (1997) Reverse genetics by chemical mutagenesis in *Caenorhabditis elegans*. Nature Gen 17:119–121

Jansen G, Thijssen KL, Werner P, van der Horst M, Hazendonk E, Plasterk RHA (1999) The complete family of genes encoding G proteins of *Caenorhabditis elegans*. Nature Gen 21:414–419

Kaplan JM, Horvitz HR (1993) A dual mechanosensory and chemosensory neuron in *Caenorhabditis elegans*. Proc Natl Acad Sci USA 90:2227–2231

Korswagen HC, Park J-H, Ohshima Y, Plasterk RHA (1997) An activating mutation in a *Caenorhabditis elegans* G_s protein induces neuronal degeneration. Genes Dev 11:1493–1503

Lochrie MA, Mendel JE, Sternberg PW, Simon MI (1991) Homologous and unique G protein alpha subunits in the nematode *Caenorhabditis elegans*. Cell Regul 2:135–154

Mello CC, Fire A (1995) DNA transformation. Methods Cell Biol 34:451–482

Mendel JE, Korswagen HC, Liu KS, Hajdu-Cronin YM, Simon MI, Plasterk RHA, Sternberg PW (1995) Participation of the G_o protein in multiple aspects of behavior in *C. elegans*. Science 267:1652–1655

Neer EJ (1995) Heterotrimeric G proteins: organizers of transmembrane signals. Cell 80:249–257

Park J-H, Ohshima S, Tani T, Ohshima Y (1997) Structure and expression of the *gsa-1* gene encoding a G protein α(s) subunit in *C. elegans*. Gene 194:183–190

Perkins LA, Hedgecock EM, Thomson JN, Culotti JG (1986) Mutant sensory cilia in the nematode *Caenorhabditis elegans*. Dev Biol 117:456–487

Roayaie K, Gage Crump J, Sagasti A, Bargmann CI (1998) The $G\alpha$ protein ODR-3 mediates olfactory and nociceptive function and controls cilium morphogenesis in *C. elegans* olfactory neurons. Neuron 20:55–67

Ségalat L, Elkes DA, Kaplan JM (1995) Modulation of serotonin controlled behaviors by G_o in *Caenorhabditis elegans*. Science 267:1648–1651

Sengupta P, Chou JH, Bargmann CI (1994) *odr-10* Encodes a seven transmembrane domain olfactory receptor required for responses to the odorant diacetyl. Cell 84:899–909

The *C. elegans* Sequencing Consortium (1998) Genome sequence of the nematode *C. elegans*: a platform for investigating biology. Science 282:2012–2018

Troemel E.R, Chou JH, Dwyer ND, Colbert HA, Bargmann CI (1995) Divergent seven transmembrane receptors are candidate chemosensory receptors in *C. elegans*. Cell 83:207–218

Troemel ER, Kimmel BE, Bargmann CI (1997) Reprogramming chemotaxis responses: sensory neurons define olfactory preferences in *C. elegans*. Cell 91:161–169

van der Voorn L, Gebbink M, Ploegh HL (1990) Characterization of a G protein β-subunit from the nematode *Caenorhabditis elegans*. J Mol Biol 213:17–26

Ward S (1973) Chemotaxis by the nematode *Caenorhabditis elegans*: Identification of attractants and analysis of the response by use of mutants. Proc Natl Acad Sci USA 70:817–821

White JG, Southgate E, Thomson JN, Brenner S (1986) The structure of the nervous system of the nematode *Caenorhabditis elegans*. Philos Trans R Soc Lond B Biol Sci314:1–340

Yu S, Avery L, Baude E, Garbers DL (1997) Guanylyl cyclase expression in specific sensory neurons: a new family of chemosensory receptors. Proc Natl Acad Sci USA 94:3384–3387

Zwaal RR, Broeks A, van Meurs J, Groenen JTM, Plasterk RHA (1993) Target selected gene inactivation in by using a frozen transposon insertion mutant bank. Proc Natl Acad Sci USA 90:7431–7435

Zwaal RR, Ahringer J, van Luenen HGAM, Rushfort A, Anderson P, Plasterk RHA (1996) G proteins are required for spatial orientation of early cell cleavage in *C. elegans* embryos. Cell 86:1–20

Zwaal RR, Mendel JE, Sternberg PW, Plasterk RHA (1997) Two neuronal G proteins are involved in the chemosensation of dauer inducing pheromone by *C. elegans*. Genetics 145:715–727

3 Programmed Cell Death and Its Regulation and Initiation in C. elegans

B. Conradt

3.1 Programmed Cell Death Is an Important Physiological Process

The proliferation of cells is an integral part of development and tissue homeostasis in multicellular animals (reviewed by Raff 1996; Follette and O'Farrell 1997). Two opposing processes, the division of cells on one hand and the programmed death of cells on the other hand, determine the overall rate of cell proliferation. The proper regulation of these two physiological processes is therefore a crucial aspect of development and of tissue homeostasis (reviewed by Edgar and Lehner 1996; Sherr 1996; Jacobson et al. 1997; Rinkenberger and Korsmeyer 1997). While the importance of the process of cell division has long been recognized, the role and extent of programmed cell death, or apoptosis, has only been realized within the last decades (Glücksmann 1950; Kerr et al. 1972). Massive programmed cell death occurs, for instance, during the development of the nervous system and in the immune system: more than 50% of all neurons and oligodendrocytes formed in the peripheral and central vertebrate nervous system undergo programmed cell death during neurogenesis (reviewed by Oppenheimer 1991; Pettmann and Henderson 1998) and more than 95% of all thymocytes generated die by programmed cell death during maturation (reviewed by Duvall and Wyllie 1986; Nagata 1997). The importance of the programmed cell-death process is underlined by the fact that when deregulated in humans, programmed cell death can cause disease (reviewed by Thompson 1995). A block in programmed cell death can lead to cancer or autoimmune diseases and increased programmed cell death appears to be involved in a number of neurodegenerative diseases.

Studies of the nematode *Caenorhabditis elegans* have contributed greatly to our current knowledge of programmed cell death. In this chapter, I will review genetic analyses in *C. elegans* that led to the identification of the central pathway required for programmed cell death in this organism. I will then summarize data which have revealed some of the molecular mechanisms that underlie this pathway. Finally, I will discuss recent findings which have implications for how the cell-death pathway might be initiated and regulated during the development and throughout the adult life of *C. elegans*.

3.2 *C. elegans* Is Amenable to Genetic Analyses of the Programmed Cell-Death Process

Due to its short life cycle, its small genome, and the ease with which it can be cultivated in large numbers, *C. elegans* has proven to be an excellent organism for genetic analyses (Wood et al. 1988; Riddle et al.

1997). The sequence of its genome has just been completed and represents the first sequenced genome of a multicellular organism (The *C. elegans* Sequencing Consortium 1998). Furthermore, the analysis of complex biological processes in *C. elegans* is greatly facilitated by the fact that the development of this organism occurs in an essentially invariant manner. The availability of the complete embryonic and postembryonic somatic cell lineage of *C. elegans* (Sulston and Horvitz 1977; Sulston et al. 1983) allows an analysis of various processes at single cell resolution.

During the development of a *C. elegans* hermaphrodite a total of 1090 somatic cells are formed. Exactly 131 out of these 1090 cells are eliminated again by a process which, at the morphological level, is reminiscent of the programmed cell-death process observed in higher organisms: the dying cells round up and shrink, the cytoplasm and chromatin condense and eventually the cell corpses are engulfed and phagocytosed by neighboring cells (Robertson and Thomson 1982). These programmed cell-death events are determined by cell lineage and occur in a cell-autonomous manner (Sulston and Horvitz 1977; Sulston et al. 1983). Programmed cell death in the *C. elegans* soma can therefore be regarded as a form of terminal differentiation or as a cell fate. Having the knowledge of which cells undergo programmed cell death in *C. elegans* and when and where this happens, allows a systematic analysis of the cell-death process at the genetic level in this organism.

3.3 A Genetic Pathway Underlies the Process of Programmed Cell Death

Using a genetic approach, mutations have been isolated that block most if not all somatic programmed cell-death events in *C. elegans* (reviewed by Ellis et al. 1991b; Metzstein et al. 1998). Animals carrying these mutations have up to 131 extra somatic cells but otherwise develop into healthy, fertile adults. The identification of cell-death-defective or *ced* mutants in *C. elegans* demonstrated for the first time that the process of programmed cell death has a genetic basis and that it is the result of the action of a set of specific genes.

The genetically identified mutations define three genes, *ced-3*, *ced-4*, and *ced-9*. Loss-of-function mutations in *ced-3* or *ced-4* block pro-

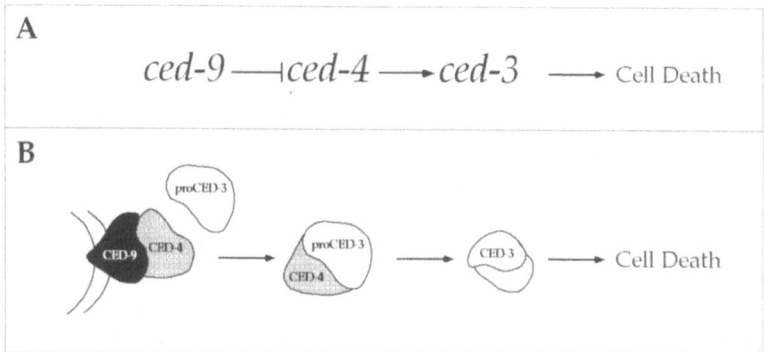

Fig. 1A,B. The central programmed cell-death pathway in *C. elegans*. **A** Genetically, the genes *ced-9*, *ced-4*, and *ced-3* appear to act in a linear pathway in which *ced-9* negatively regulates *ced-4*, and *ced-4* positively regulates *ced-3* which is required for programmed cell death. **B** The gene products of *ced-9*, *ced-4*, and *ced-3* interact physically (see text for details)

grammed cell death, indicating that their gene products are required for killing (Ellis and Horvitz 1986). In contrast, *ced-9* encodes a cell-death inhibitor: while a gain-of-function mutation in *ced-9* blocks programmed cell death, loss-of-function mutations in *ced-9* cause ectopic programmed cell death resulting in embryonic lethality (Hengartner et al. 1992). This lethality can be suppressed by loss-of-function mutations in either *ced-3* or *ced-4*, suggesting that *ced-3* and *ced-4* act downstream of, or in parallel to, *ced-9* and that *ced-9* normally acts to block *ced-3* and *ced-4*. Genetic mosaic analyses indicate that *ced-3* and *ced-4* are required in cells destined to die (Yuan and Horvitz 1990) which suggests that both genes act in a cell-autonomous manner. Furthermore, the overexpression of either *ced-3* or *ced-4* in *C. elegans* in cells that normally survive can induce these cells to undergo programmed cell death (Shaham and Horvitz 1996a). The overexpression of *ced-3* and *ced-4* in *C. elegans* also allowed to order these two genes with respect to each other. While *ced-4*-induced killing occurs only in animals that have a functional *ced-3* gene, *ced-3*-induced killing can occur in animals that carry a loss-of-function mutation in *ced-4* (Shaham and Horvitz 1996a). Furthermore, the ability of *ced-9* to block *ced-3*-induced killing depends on a functional *ced-4* gene (Shaham and Horvitz 1996a). These observations suggest that *ced-4* most likely acts upstream of, or in parallel to, *ced-3*. *ced-9*, *ced-4*, and *ced-3* therefore appear to act in a simple pathway in which *ced-9* negatively regulates *ced-4* and *ced-4* positively regulates *ced-3* (Fig. 1A).

The programmed cell-death pathway, however, is probably more complex. *ced-9* loss-of-function mutations, for instance, enhance the cell-death defect caused by weak loss-of-function mutations in either *ced-3* or *ced-4*, suggesting that *ced-9* not only has a cell-death protective function but also a killing function (Hengartner and Horvitz 1994b; Conradt and Horvitz 1998). In addition, the gene products of *ced-9* and *ced-3* might interact directly (see below). Also, the *ced-4* message is alternatively spliced (Shaham and Horvitz 1996b). This results in the production of the major "short" transcript, *ced-4*s, and a minor "long" transcript, *ced-4*l, which contains an additional 72 nucleotides in the center of the message. Whereas the overexpression of the *ced-4*s message induces killing in *C. elegans*, the overexpressing of *ced-4*l blocks programmed cell death (Shaham and Horvitz 1996b).

3.4 The Programmed Cell-Death Pathway Has Been Conserved Through Evolution

The cloning of the *ced-9* gene confirmed that the process of programmed cell death is conserved not only at the cellular level but also at the molecular level (Hengartner and Horvitz 1994a). *ced-9* was found to encode a protein with structural and functional similarity to the mammalian oncoprotein and cell-death inhibitor Bcl-2, the prototype of the growing family of Bcl-2-like cell-death regulators (reviewed by Chao and Korsmeyer 1998; Adams and Cory 1998). The *bcl-2* gene was originally identified at the breakpoint of the t(14;18) translocation present in the majority of individuals with follicular lymphomas. The t(14;18) translocation results in the inappropriate overexpression of the *bcl-2* gene in B cells and the extended survival of these cells. It has since been shown that the overexpression of *bcl-2* in a number of cell types prolongs survival by blocking programmed cell death (Chao and Korsmeyer 1998; Adams and Cory 1998). The cloning of the *ced-3* gene revealed a role for proteolysis in the cell-death process. *ced-3* encodes a cysteine protease similar to ICE, the interleukin-1β-converting enzyme (Yuan et al. 1993). A large number of CED-3/ICE-like proteases with various functions in programmed cell death have since been identified in both vertebrates and invertebrates and are now collectively called caspases (reviewed by Nicholson and Thornberry 1997; Cryns and Yuan 1998). Finally, a mammalian protein identified biochemically as required for programmed cell death in vitro, Apaf-1 (Zhou et al. 1997), shows similarity to the product of the *ced-4* gene (Yuan and Horvitz 1992).

The importance of the mammalian counterparts of *ced-9*, *ced-3*, and *ced-4* in the programmed cell-death process has been confirmed by the phenotypic analysis of mice carrying targeted mutations in the respective genes. Mice lacking functional copies of the *bcl-2* or *bcl-x* gene, which encode two Bcl-2-like cell-death inhibitors, exhibit developmental defects of varying degrees which can be attributed to increased programmed cell death. While *bcl-2*[−/−] mice exhibit a defect in kidney development, significantly lower numbers of a subset of cranial sensory neurons at the time of birth, and increased loss of sensory, sympathetic, and motor neurons postnatally (Veis et al. 1993; Michaelidis et al. 1996; Piñón et al. 1997), Bcl-x deficient mice show signs of massive ectopic cell death of immature hematopoietic cells and of neurons in the brain, in the spinal cord and in the dorsal root ganglia, resulting in embryonic lethality (Motoyama et al. 1995). Mice deficient in Caspases 3 or 9, or in Apaf-1, in contrast, have defects indicative of a lack of programmed cell death. In all three cases, brain development is severely affected due to reduced levels of programmed cell death resulting in hyperplasia, in disorganized brain development, and subsequently in abnormal skull formation, resulting in lethality (Kuida et al. 1996; Woo et al. 1998; Kuida et al. 1998; Hakem et al. 1998; Cecconi et al. 1998; Yoshida et al. 1998). *Apaf-1*[−/−] mice, in addition, exhibit defects in the sculpting of the limbs and in the formation of the palate (Cecconi et al. 1998; Yoshida et al. 1998), two processes known to require programmed cell death (Garcia-Martinez et al. 1993; Ferguson 1988; reviewed by Coucouvanis et al. 1995).

3.5 Elucidating the Molecular Mechanism of the Programmed Cell-Death Process

The genes *ced-9*, *ced-4*, and *ced-3* form the central programmed cell-death pathway in *C. elegans* and so far, there is no evidence of additional *ced-9-*, *ced-4-*, or *ced-3*-like genes with a role in programmed cell death in this organism. This is in contrast to vertebrates in which a family of *ced-9*-like genes exists (i.e., the family of *bcl-2*-like genes) and in which over ten CED-3/ICE-like cell-death proteases have been identified. The *C. elegans* cell-death pathway can therefore be considered the "basic model" for the programmed cell-death process.

Using various biochemical analyses, the yeast two-hybrid system, and the expression of the CED-9, CED-4, and CED-3 proteins in heterologous systems, such as mammalian or insect cells, data have been gathered over the last few years which reveal some aspects of the molecular mechanism of the cell-death process. While most of the results still need to be confirmed in *C. elegans*, the following model is emerging from these results (Fig. 1B). The protease activity of the CED-3 caspase is required for killing (Xue et al. 1996; Seshagiri and Miller 1997; Chinnaiyan et al. 1997a; Wu et al. 1997b; Hisahara et al. 1998). Because CED-3, just like other caspases, is synthesized as an inactive zymogen, proCED-3, the activation of CED-3 is an important step in the cell-death process. It has been shown that caspase zymogens can be activated through auto-catalysis (reviewed by Nicholson and Thornberry 1997; Cryns and Yuan 1998). However, the cleavage of proCED-3 to the mature, active CED-3 caspase is greatly enhanced by the CED-4 protein (Seshagiri and Miller 1997; Wu et al. 1997b; Chinnaiyan et al. 1997b) which can interact with proCED-3 (Chinnaiyan et al. 1997a; Wu et al. 1997b; Irmler et al. 1997). This interaction appears to be mediated through two distinct domains of CED-4: an N-terminal domain of CED-4 interacts with the catalytic domain of proCED-3 and the P-loop region of CED-4, a nucleotide-binding domain, interacts with the pro-domain of proCED-3 (Chaudhary et al. 1998). The enhancement of proCED-3 cleavage and activation by the CED-4 protein has been proposed to come about through the oligomerization of CED-4 which would allow bound proCED-3 molecules to cleave and activate each other (Yang et al. 1998). It has also been suggested that the interaction between the P-loop region of CED-4 and the pro-domain of proCED-3 is required for proCED-3 activation (Chaudhary et al. 1998). Interestingly, the insertion into the P-loop region of 24 amino acids in the CED-4L protein results in the disruption of this specific interaction and the inability of the CED-4L protein to activate proCED-3 (Chaudhary et al. 1998). CED-4 also interacts with the cell-death inhibitor protein CED-9 (Spector et al. 1997; Chinnaiyan et al. 1997a; Wu et al. 1997a; Seshagiri and Miller 1997; Ottilie et al. 1997), the latter being membrane-associated due to a hydrophobic region at the C-terminus (Wu et al. 1997a; Chaudhary et al. 1998; Hisahara et al. 1998). The presence of CED-9 blocks the CED-4-dependent activation of proCED-3 (Chinnaiyan et al. 1997a; Seshagiri and Miller 1997; Wu et al. 1997b). It is therefore likely that, by binding to CED-4, CED-9 inhibits the ability of CED-4 to activate proCED-3. CED-9 might block the oligomerization of CED-4 or disrupt the interaction of the P-loop region of CED-4 with proCED-3 (Yang et al. 1998; Chaudhary et al. 1998). CED-9, CED-4, and proCED-3 possibly form a ternary complex (Chinnaiyan et al. 1997a; Wu et al. 1997b) which has been proposed to be present in all cells during *C. elegans* development (Vaux 1997; Hengartner 1997a). The programmed death of cells would be blocked as long as the inhibitory interaction between CED-9 and CED-4 is maintained. The possibility that CED-9, CED-4, and proCED-3 are present not only in cells destined to die is supported by the following observations: loss-of-function mutations in the *ced-9* gene cause cells that normally survive to die in a *ced-3-* and *ced-4*-dependent manner (Hengartner et al. 1992); and mutations in *ced-9*, *ced-3*, or *ced-4* influence the efficiency with which the overexpression of either *ced-4* or *ced-3* in *C. elegans* can kill cells that normally survive (Shaham and Horvitz 1996a).

A number of the molecular events just described also appear to occur in vertebrates. It has been shown, for example, that Apaf-1 binds to, and is required for, the activation of Caspase 9 (Li et al. 1997). In addition, the Apaf-1-dependent activation of Caspase 9 in vitro appears to involve the oligomerization of Apaf-1 (Srinivasula et al. 1998) and is blocked by the Bcl-2-like cell-death inhibitor Bcl-x which can interact with Apaf-1 (Pan et al. 1998; Hu et al. 1998).

3.6 The Gene Product of *egl-1* Initiates Programmed Cell Death in *C. elegans*

The emerging molecular model raises a number of questions, such as what substrates of the CED-3 caspase are required to bring about cell death. While a number of potential substrates have been identified for vertebrate caspases (reviewed by Porter et al. 1997), the only *C. elegans* substrate of CED-3 characterized so far is the protein CED-9 (Xue and Horvitz 1997). It is possible that the CED-3-dependent cleavage of CED-9 is required for the progression of the cell-death process but it is also possible that, by being a substrate for CED-3, CED-9 acts as a competitive inhibitor of CED-3; this would constitute a safeguard mechanism against inappropriately activated CED-3 molecules in cells destined to survive. Another question raised by the emerging model is how programmed cell death is initiated in cells destined to die. The recent identification of another *C. elegans* gene required for programmed cell death in the soma, *egl-1* (*egl*, *egg*-laying defective), has provided some clues to the mechanism of cell-death initiation.

Just like loss-of-function mutations in *ced-3* or *ced-4*, a loss-of-function mutation in the gene *egl-1* blocks most if not all somatic cell death events (Conradt and Horvitz 1998). However, unlike loss-of-function mutations in *ced-3* or *ced-4*, the *egl-1* loss-of-function mutation fails to suppress *ced-9* loss-of-function mutations. In animals carrying both an *egl-1* loss-of-function mutation and a *ced-9* loss-of-function mutation, ectopic cell death still occurs resulting in lethality. This suggests that *egl-1* is acting upstream of, or in parallel to, *ced-9*. The *egl-1* loss-of-function mutation can enhance the cell-death defect caused by weak *ced-3* loss-of-function mutations. However, this ability of the *egl-1* loss-of-function mutation is dependent on the presence of a functional *ced-9* gene; in the absence of *ced-9*, the *egl-1* loss-of-function mutation fails to enhance a weak *ced-3* lf mutation. This indicates that *egl-1* acts through *ced-9* and that it normally functions to negatively regulate *ced-9* (Conradt and Horvitz 1998) (Fig. 2A). The overexpression of *egl-1* in *C. elegans* in cells that normally survive causes these cells to undergo programmed cell death. This indicates that, like *ced-3* and *ced-4*, the *egl-1* gene acts in a cell-autonomous manner. *egl-1*-induced killing is suppressed by the gain-of-function mutation in *ced-9* and by loss-of-function mutations in either *ced-4* or *ced-3* which provides further

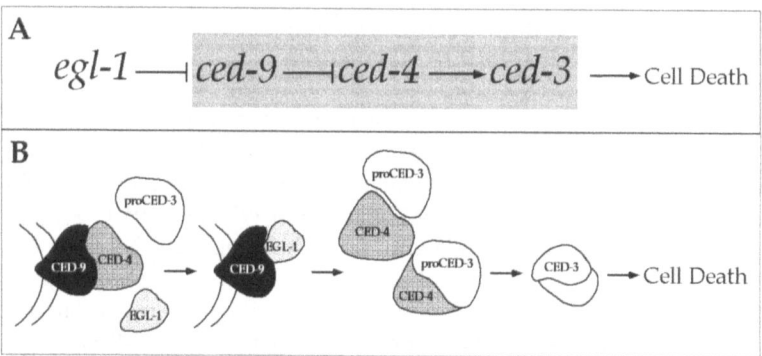

Fig. 2A,B. Initiation of programmed cell death in *C. elegans*. **A** Genetically, the *egl-1* gene initiates programmed cell death by negatively regulating *ced-9*. **B** The EGL-1 protein might initiate programmed cell death by binding to, thereby regulating, the cell-death inhibitor protein CED-9 (see text for details)

support for the model that *egl-1* acts upstream of *ced-9*, *ced-4*, and *ced-3* genetically (Conradt and Horvitz 1998).

The *egl-1* gene encodes a small protein with a domain that is reminiscent of the Bcl-2-homology region 3, called BH3 domain (Conradt and Horvitz 1998). The EGL-1 protein is therefore similar in structure to the members of a rapidly growing subfamily of the family of Bcl-2-like cell-death regulators, the BH3-only containing proteins, which include the mammalian proteins Bad, Bid, Blk, Hrk/DP5, Bim, and NIP3 (reviewed by Adams and Cory 1998; Kelekar and Thompson 1997). The BH3-only containing proteins were initially identified by virtue of their binding to Bcl-2-like proteins. This binding depends on the BH3 domain of the BH3-only containing proteins which interacts with a hydrophobic cleft formed by the BH1, BH2, and BH3 domain of the Bcl-2-like proteins. When overexpressed in mammalian cells, BH3-only containing proteins can induce programmed cell death, and their ability to do so appears to depend on their ability to bind to Bcl-2-like proteins. However, the physiological role of these proteins in the programmed cell-death process has not been established yet. The fact that *egl-1* acts upstream of *ced-9* genetically, suggests that at least some of the mammalian BH3-only containing proteins might similarly act as negative regulators of Bcl-2-like cell-death inhibitors.

As expected from results obtained with the mammalian BH3-only containing proteins, the EGL-1 protein interacts with the Bcl-2-like cell-death inhibitor CED-9 in a BH3-dependent manner in vitro and in a yeast two-hybrid system (Conradt and Horvitz 1998). This finding suggests a molecular mechanism for the initiation of the programmed cell-death process in *C. elegans* (Fig. 2B). *egl-1* initiates programmed cell death in *C. elegans* in a *ced-4-* and *ced-3*-dependent manner. Furthermore, *egl-1* acts through *ced-9*. It is therefore likely that the EGL-1 protein induces programmed cell death by binding to the CED-9 protein, thereby negatively regulating the ability of CED-9 to inhibit CED-4. How the binding of EGL-1 to CED-9 might negatively regulate CED-9 is not known. However, it has been suggested that the binding of EGL-1 to CED-9 results in the release of CED-4 from the membrane-bound complex (del Peso et al. 1998) (Fig. 2B). CED-4 could, for example, be displaced if EGL-1 and CED-4 competed for the same binding site or if the binding of EGL-1 to CED-9 resulted in a conformational change in the CED-9 protein which disrupts the interaction between CED-9 and CED-4.

3.7 Cell-Death Initiation Might Be Regulated by Cell-Type Specific Factors

The EGL-1 protein can be regarded as the initiator of programmed cell death in the *C. elegans* soma. But how is EGL-1 itself regulated? The mammalian BH3-only containing proteins appear to be regulated by diverse mechanisms. The removal of nerve growth factor from cultures of sympathetic neurons, for instance, results in the transcriptional upregulation of the BH3-only containing protein DP5 (Imaizumi et al. 1997). The availability of growth factors also regulates the activity of the BH3-only containing protein Bad. The ability of Bad to bind to Bcl-2-like cell-death inhibitors and to induce programmed cell death, however, is regulated at the posttranslational level through the phosphorylation of Bad at a critical serine residue (reviewed by Franke and Cantley 1997). The BH3-only containing protein Bid is also regulated at the posttranslational level, in this case through proteolytic cleavage at specific sites (Luo et al. 1998; Li et al. 1998).

CENTRAL PATHWAY

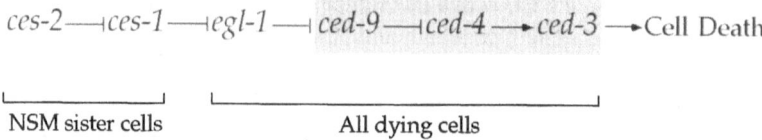

NSM sister cells	All dying cells

Fig. 3. Regulation of programmed cell death in the *C. elegans* soma by cell type-specific factors. The genes *ces-2* and *ces-1* act as regulators of programmed cell death in the NSM sister cells which normally undergo programmed cell death (see text for details)

The *C. elegans* genes *ces-1* and *ces-2* (*ces*, cell-death specification) provide some hints about how the initiation of programmed cell death might be regulated in the *C. elegans* soma (Ellis and Horvitz 1991a). Unlike mutations in *egl-1*, *ced-9*, *ced-4*, and *ced-3*, which can block most if not all somatic cell death events, mutations in *ces-1* and *ces-2* only block a subset of these cell deaths. Gain-of-function mutations in *ces-1* and loss-of-function mutations in *ces-2* block the deaths of two specific neurons, the NSM sister cells, which normally undergo programmed cell death. In animals carrying a *ces-1* loss-of-function mutation, the NSM sister cells die, as they do in wild-type animals, and they also die in animals carrying both a loss-of-function mutation in *ces-1* and a loss-of-function mutation in *ces-2*. This suggests that *ces-1* acts downstream of, or in parallel to, *ces-2*. Furthermore, in animals carrying both a *ces-1* loss-of-function mutation and a loss-of-function mutation in either *egl-1*, *ced-4*, or *ced-3*, the NSM sister cells survive, indicating that *ces-1* and *ces-2* are likely to act upstream of, or in parallel to, *egl-1*, *ced-4*, and *ced-3* (Ellis and Horvitz 1991a; Conradt and Horvitz 1998) (Fig. 3). This makes *ces-2* and *ces-1* good candidates for NSM sister cell-specific regulators of the central cell-death pathway, and of *egl-1* in particular, due to its position in the pathway. The phenotype of the *ces-1* and *ces-2* mutations furthermore suggests that the life-versus-death decision of individual cells is regulated by cell type-specific factors, such as the gene products of *ces-2* and *ces-1*, and that there are probably many *ces-2* and *ces-1*-like components yet to be discovered which are involved in regulating the death of all the 131 somatic cells that undergo programmed cell death during *C. elegans* development.

The *ces-2* gene has been cloned and encodes a basic region leucine-zipper (bZIP) transcription factor most similar in its DNA-binding and dimerization domain to the human hepatic leukemia factor, HLF (Metzstein et al. 1996). An oncogenic form of HLF found in a particular form of pro-B-cell cancer (acute lymphoblastic leukemia), the E2A-HLF fusion protein, is composed of the transactivation domain of E2A and the DNA-binding domain of HLF and blocks the programmed cell death of pro-B-cells (Inaba et al. 1996). It has been proposed that the E2A-HLF fusion protein blocks programmed cell death by competing with a CES-2-like factor for target sites in the promoter of a cell-death promoting gene (Inaba et al. 1996; Metzstein et al. 1996). This finding suggests that there might be conservation between *C. elegans* and mammals not only within the central cell-death pathway but also between the components that regulate this pathway. No target gene of the CES-2 transcription factor has been characterized yet but it is likely that *ces-1*, which acts downstream of *ces-2* genetically, represents such a target.

One additional cell type-specific regulator of the central cell-death pathway might have been discovered with the identification of gain-of-function mutations in *egl-1*. These gain-of-function mutations were identified in screens for egg-laying defective mutants and initially defined the *egl-1* gene (Trent et al. 1983; Desai and Horvitz 1989). Gain-of-function mutations in *egl-1* cause the inappropriate programmed cell death of a pair of serotonergic motor neurons, the HSNs or hermaphrodite-specific neurons, which are required for egg laying in hermaphrodites (Sulston and Horvitz 1977). The phenotype of these gain-of-function mutations suggests that they affect the regulation of the *egl-1* gene by a HSN-specific regulator of programmed cell death.

The identification of the *ces-1* gene, as well as the nature of the *egl-1* gain-of-function mutations, has not been reported yet.

3.8 Programmed Cell Death Is Initiated
by Two Independent Pathways in *C. elegans*

Programmed cell death in *C. elegans* occurs not only in somatic tissues during *C. elegans* embryonic and postembryonic development but also in the germ line of adult hermaphrodites (Sulston 1988; Hengartner 1997b). A large number of germ cells, possibly more than 50% of all germ cells formed, undergo programmed cell death. The number of cell-death events in this tissue appears to increase with the age of the animal and seems to depend on the availability of food which suggests that the programmed cell-death process in the germ line might be regulated or at least influenced by extra-cellular, cell non-autonomous signals (Sulston 1988).

In hermaphrodites carrying loss-of-function mutations in either *ced-4* or *ced-3*, programmed cell death in the germ line is blocked (Hengartner 1997b). This suggests that *ced-4* and *ced-3* are required for killing not only in the soma but also in the germ line. Loss-of-function mutations in *ced-9* cause ectopic cell death resulting in embryonic lethality. However, *ced-9* loss-of-function progeny of hermaphrodites that are heterozygous for a *ced-9* loss-of-function mutation are viable due to maternal rescue. These *ced-9* loss-of-function animals develop into adults with low fertility as a result of the increased programmed cell death of germ cells. This suggests that *ced-9* acts as a cell-death inhibitor also in the germ line (Hengartner 1997b). The gain-of-function mutation in *ced-9*, which blocks programmed cell death in the soma, however, does not block programmed cell death in the germ line. This suggests that differences exist in the way that programmed cell death is initiated in the soma and in the germ line. This is supported by the finding that the *egl-1* gene, which is required for programmed cell death in the soma just like *ced-4* and *ced-3* are, is not involved in programmed cell death in the germ line (Conradt and Horvitz 1998).

egl-1, hence, is not the only initiator of programmed cell death in *C. elegans*. Instead, two independent pathways appear to initiate programmed cell death in this organism: an *egl-1*-dependent pathway that functions in the soma and an *egl-1*-independent pathway that acts in the germ line (Fig. 4). So far, there are no hints as to the nature of the *egl-1*-independent pathway except that it might involve the transduction of an extracellular signal. Considering, however, that *ced-9* is also

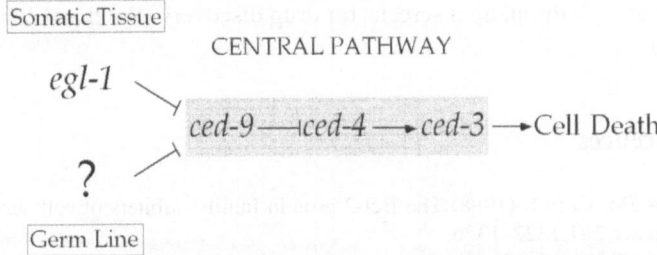

Fig. 4. Programmed cell death in *C. elegans* is regulated and initiated by two independent processes: an *egl-1*-dependent process, which functions in the soma during *C elegans* development; and an *egl-1*-independent process, which acts in the germ line of adult hermaphrodites (see text for details)

involved in programmed cell death in the germ line, we speculate that the *ced-9* gene or the CED-9 protein might be the target of this *egl-1*-independent pathway (Fig. 4).

3.9 Conclusions

Genetic analyses in *C. elegans* have contributed greatly to our current knowledge of the biology of the programmed cell-death process. However, many questions remain about how this process works at the molecular level, how it brings about the death of cells, and how this process is initiated and regulated. Recent genetic and molecular analyses in *C. elegans* have given us more insight into how programmed cell death is initiated and regulated in this organism and suggest that conservation between species might exist even at this level of the cell-death process. The dissection of the mechanisms and players involved in the regulation and initiation of the programmed cell-death process in *C. elegans* is therefore likely to provide information that will be applicable to other organisms as well.

C. elegans might be a useful tool not only in the quest to understand the biology of the programmed cell-death process but also to identify potential inhibitors and activators of this process. The fact that large numbers of animals can efficiently be analyzed makes *C. elegans* amenable to high throughput screens for drug discovery (Rand and Johnson 1995).

References

Adams JM, Cory S (1998) The Bcl-2 protein family: arbiters of cell survival. Science 281:1322–1326

Cecconi F, Alvarez-Bolado G, Meyer BI, Roth KA, Gruss P (1998) Apaf-1 (CED-4 homolog) regulates programmed cell death in mammalian development. Cell 94:727–737

Chao DT, Korsmeyer SJ (1998) Bcl-2 family: regulators of cell death. Annu Rev Immunol 16:395–419

Chaudhary D, O'Rourke K, Chinnaiyan AM, Dixit VM (1998) The death inhibitory molecules CED-9 and CED-4L use a common mechanism to inhibit the CED-3 death protease. J Biol Chem 273:17708–17712

Chinnaiyan AM, O'Rourke K, Lane BR, Dixit VM (1997a) Interaction of CED-4 with CED-3 and CED-9: a molecular framework for cell death. Science 275:1122–1126

Chinnaiyan AM, Chaudhary D, O'Rourke K, Koonin EV, Dixit VM (1997b) Role of CED-4 in the activation of CED-3. Nature 388:728–729

Conradt B, Horvitz HR (1998) The *C. elegans* protein EGL-1 is required for programmed cell death and interacts with the Bcl-2-like Protein CED-9. Cell 93:519–529

Coucouvanis EC, Martin GR, Nadeau JH (1995) Genetic approaches for studying programmed cell death during development of the laboratory mouse. Methods Cell Biol 46:387–440

Cryns V, Yuan J (1998) Proteases to die for. Genes Dev 12:1551–1570

del Peso L, González VM, Núñez G (1998) *Caenorhabditis elegans* EGL-1 disrupts the interaction of CED-9 with CED-4 and promotes CED-3 activation. J Biol Chem 273:33495–33500

Desai C, Horvitz HR (1989) *Caenorhabditis elegans* mutants defective in the functioning of the motor neurons responsible for egg laying. Genetics 121:703–721

Duvall E, Wyllie AH (1986) Death and the cell. Immunol Today 7:115–119

Edgar BA, Lehner CF (1996) Developmental control of cell cycle regulators: a fly's perspective. Science 274:1646–1652

Ellis HM, Horvitz HR (1986) Genetic control of programmed cell death in the nematode *C. elegans*. Cell 44:817–829

Ellis RE, Horvitz HR (1991a) Two *C. elegans* genes control the programmed deaths of specific cells in the pharynx. Development 112:591–603

Ellis RE, Yuan J, Horvitz HR (1991b) Mechanisms and functions of cell death. Annu Rev Cell Biol 7:663–698

Ferguson MW (1988) Palate development. Dev Suppl 103:41–60

Follette PJ, O'Farrell PH (1997) Connecting cell behavior to patterning: lessons from the cell cycle. Cell 88:309–314

Franke TF, Cantley LC (1997) A Bad kinase makes good. Nature 390:116–117

Garcia-Martinez V, Macias D, Gañan Y, Garcia-Lobo JM, Francia MV, Fernandez-Teran MA, Hurle JM (1993) Internucleosomal DNA fragmentation and programmed cell death (apoptosis) in the interdigital tissue of the embryonic chicken leg bud. J Cell Sci 106:201–208

Glücksmann A (1950) Cell deaths in normal vertebrate ontogeny. Biol Rev Camb Philos Soc 26:59–86

Hakem R, Hakem A, Duncan GS, Henderson JT, Woo M, Soengas MS, Elia A, de la Pompa JL, Kagi D, Khoo W, Potter J, Yoshida R, Kaufman SA, Lowe SW, Penninger JM, Mak TW (1998) Differential requirement for Caspase 9 in apoptotic pathways in vivo. Cell 94:339–352

Hengartner MO (1997a) CED-4 is a stranger no more. Nature 388:714–715

Hengartner MO (1997b) Cell death. In: Riddle DL, Blumenthal T, Meyer BJ, Priess JR (eds) *C. elegans* II. Cold Spring Harbor Laboratory Press, New York, p 383

Hengartner MO, Ellis RE, Horvitz HR (1992) *Caenorhabditis elegans* gene *ced-9* protects cells from programmed cell death. Nature 356:494–499

Hengartner MO, Horvitz HR (1994a) *C. elegans* cell survival gene *ced-9* encodes a functional homolog of the mammalian proto-oncogene *bcl-2*. Cell 76:665–676

Hengartner MO, Horvitz HR (1994b) Activation of *C. elegans* cell death protein CED-9 by an amino-acid substitution in a domain conserved in Bcl-2. Nature 369:318–320

Hisahara S, Kanuka H, Shoji S-i, Yoshikawa S, Okano H, Miura M (1998) *Caenorhabditis elegans* anti-apoptotic gene *ced-9* prevents *ced-3*-induced cell death in *Drosophila* cells. J Cell Sci 111:667–673

Hu Y, Benedict MA, Wu D, Inohara N, Núñez G (1998) Bcl-X_L interacts with Apaf-1 and inhibits Apaf-1-dependent caspase-9 activation. Proc Natl Acad Sci USA 95:4386–4391

Imaizumi K, Tsuda M, Imai Y, Wanaka A, Takagi T, Tohyama M (1997) Molecular cloning of a novel polypeptide, DP5, induced during programmed neuronal death. J Biol Chem 272:18842–18848

Inaba T, Inukai T, Yoshihara T, Seyschab H, Ashmun RA, Canman CE, Laken SJ, Kastan MB, Look AT (1996) Reversal of apoptosis by the leukaemia-associated E2A-HLF chimeric transcription factor. Nature 382:541–544

Irmler M, Hofmann K, Vaux D, Tschopp J (1997) Direct physical interaction between the *Caenorhabditis elegans* 'death proteins' CED-3 and CED-4. FEBS Lett 406:189–190

Jacobson MD, Weil M, Raff MC (1997) Programmed cell death in animal development. Cell 88:347–354

Kelekar A, Thompson CB (1997) Bcl-2 family proteins: the role of the BH3 domain in apoptosis. Trends Cell Biol 8:324–330

Kerr JFR, Wyllie AH, Currie AR (1972) Apoptosis: a basic biological phenomenon with wide-range implications in tissue kinetics. Br J Cancer 26:239–257

Kuida K, Zheng TS, Na S, Kuan C-yK, Yang D, Karasuyama H, Rakic P, Flavell RA (1996) Decreased apoptosis in the brain and premature lethality in CPP-32 deficient mice. Nature 384:368–372

Kuida K, Haydar TF, Kuan C-Y, Gu Y, Taya C, Karasuyama H, Su MS-S, Rakic P, Flavell RA (1998) Reduced apoptosis and cytochrome c-mediated caspase activation in mice lacking Caspase 9. Cell 94:325–337

Li P, Nijhawan D, Budihardjo I, Srinivasula SM, Ahmad M, Alnemri ES (1997) Cytochrome c and dATP-dependent formation of Apaf-1/Caspase-9 complex initiates an apoptotic protease cascade. Cell 91:479–489

Li H, Zhu H, Xu C-j, Yuan J (1998) Cleavage of BID by caspase 8 mediates the mitochondrial damage in the Fas pathway of apoptosis. Cell 94:491–501

Luo X, Budihardjo I, Zou H, Slaughter C, Wang X (1998) Bid, a Bcl2 interacting protein, mediates cytochrome c release from mitochondria in response to activation of cell surface death receptors. Cell 94:481–490

Metzstein MM, Hengartner MO, Tsung N, Ellis RE, Horvitz HR (1996) Transcriptional regulator of programmed cell death encoded by *Caenorhabditis elegans* gene *ces-2*. Nature 328:545–547

Metzstein MM, Stanfield GS, Horvitz HR (1998) Genetics of programmed cell death in *C. elegans*: past, present and future. Trends Genet 14:410–416

Michaelidis TM, Sendtner M, Cooper JD, Airaksinen MS, Holtmann B, Meyer M, Thoenen H (1996) Inactivation of *bcl-2* results in progressive degeneration of motoneurons, sympathetic and sensory neurons during early postnatal development. Neuron 17:75–89

Motoyama N, Wang F, Roth KA, Sawa H, Nakayama K-i, Nakayama K, Negishi I, Senju S, Zhang Q, Fujii S, Loh DY (1995) Massive cell death of immature hematopoietic cells and neurons in Bcl-x-deficient mice. Science 267:1506–1510

Nagata S (1997) Apoptosis by death factor. Cell 88:355–365

Nicholson DW, Thornberry NA (1997) Caspases: killer proteases. Trends Biochem Sci 22:299–306

Oppenheimer RW (1991) Cell death during development of the nervous system. Annu Rev Neurosci 14:453–501

Ottilie S, Wang Y, Banks S, Chang J, Vigna NJ, Weeks S, Armstrong RC, Fritz LC, Oltersdorf T (1997) Mutational analysis of the interacting cell death regulators CED-9 and CED-4. Cell Death Diff 4:526–533

Pan G, O'Rourke K, Dixit VM (1998) Caspase-9, Bcl-X_L, and Apaf-1 form a ternary complex. J Biol Chem 273:5841–5845

Pettmann B, Henderson CE (1998) Neuronal cell death. Neuron 20:633–647

Piñón LGP, Middleton G, Davies AM (1997) Bcl-2 is required for cranial sensory neuron survival at defined stages of embryonic development. Development 124:4173–4178

Porter AG, Ng P, Jänicke RU (1997) Death substrates come alive. Bioessays 19:501–507

Raff MC (1996) Size control: the regulation of cell numbers in animal development. Cell 86:173–175

Rand JB, Johnson, CD (1995) Genetic pharmacology: Interactions between drugs and gene products. Methods Cell Biol 48:187–204

Riddle DL, Blumenthal T, Meyer BJ, Priess JR (1997) *C. elegans* II, Cold Spring Harbor Laboratory Press, New York

Rinkenberger JL, Korsmeyer SJ (1997) Errors of homeostasis and deregulated apoptosis. Curr Opin Genet Dev 7:589–596

Robertson AMG, Thomson JN (1982) Morphology of programmed cell death in the ventral cord of *Caenorhabditis elegans* larvae. J Embryol Exp Morph 67:89–100

Seshagiri S, Miller LK (1997) *Caenorhabditis elegans* CED-4 stimulates CED-3 processing and CED-3-induced apoptosis. Curr Biol 7:455–460

Shaham S, Horvitz HR (1996a) Developing *Caenorhabditis elegans* neurons may contain both cell-death protective and killer activities. Genes Dev 10:578–591

Shaham S, Horvitz HR (1996b) Alternatively spliced *C. elegans ced-4* RNA encodes a novel cell death inhibitor. Cell 86:201–208

Sherr CJ (1996) Cancer cell cycles. Science 274:1672–1677

Srinivasula SM, Ahmad M, Fernandes-Alnemri T, Alnemri ES (1998) Autoactivation of procaspase-9 by Apaf-1-mediated oligomerization. Mol Cell 1:949–957

Spector MS, Desnoyers S, Hoeppner DJ, Hengartner MO (1997) Interactions between the *C. elegans* cell-death regulators CED-9 and CED-4. Nature 385:653–656

Sulston, J (1988) Cell lineage. In: Wood WB and the Community of *C. elegans* Researchers (eds) The Nematode *Caenorhabditis elegans*. Cold Spring Harbor Laboratory Press, New York, p 123

Sulston JE, Horvitz HR (1977) Post-embryonic cell lineages of the nematode *Caenorhabditis elegans*. Dev Biol 56:110–156

Sulston JE, Schierenberg E, White JG, Thomson JN (1983) The embryonic cell lineage of the nematode *Caenorhabditis elegans*. Dev Biol 100:64–119

The *C. elegans* Sequencing Consortium (1998) Genome sequence of the nematode *C. elegans*: a platform for investigating biology. Science 282:2012–2018

Thompson CB (1995) Apoptosis in the pathogenesis and treatment of diseases. Science 267:1456–1462

Trent C, Tsung N, Horvitz HR (1983) Egg-laying defective mutants of the nematode *Caenorhabditis elegans*. Genetics 104:619–647

Vaux DL (1997) CED-4 – The third horseman of apoptosis. Cell 90:389–390

Veis DJ, Sorenson C, Shutter JR, Korsmeyer SJ (1993) Bcl-2-deficient mice demonstrate fulminant lymphoid apoptosis, polycystic kidneys, and hypopigmented hair. Cell 75:229–240

Woo M, Hakem R, Soengas MS, Duncan GS, Shahinian A, Kägi D, Hakem A, McCurrach M, Khoo W, Kaufman SA, Senaldi G, Howard T, Lowe SW, Mak TW (1998) Essential contribution of Caspase 3/CPP32 to apoptosis and its associated nuclear changes. Genes Dev 12:806–819

Wood WB and the community of *C. elegans* researchers (1988) The nematode *Caenorhabditis elegans*, Cold Spring Harbor Laboratory Press, New York

Wu D, Wallen HD, Núñez G (1997a) Interaction and regulation of subcellular localization of CED-4 and CED-9. Science 275:1126–1129

Wu D, Wallen HD, Inohara N, Núñez G (1997b) Interaction and regulation of the *Caenorhabditis elegans* death protease CED-3 by CED-4 and CED-9. J Biol Chem 272:21449–21454

Xue D, Horvitz HR (1997) *Caenorhabditis elegans* CED-9 protein is a bifunctional cell-death inhibitor. Nature 390:305–308

Xue D, Shaham S, Horvitz HR (1996) The *Caenorhabditis elegans* cell-death protein CED-3 is a cysteine protease with substrate specificities similar to those of the human CPP32 protease. Genes Dev 10:1073–1083

Yang X, Chang HY, Baltimore D (1998) Essential role of CED-4 oligomerization in CED-3 activation and apoptosis. Science 281:1355–1357

Yoshida H, Kong Y-Y, Yoshida R, Elia AJ, Hakem A, Hakem R, Penninger JM, Mak TW (1998) Apaf-1 is required for mitochondrial pathways of apoptosis and brain development. Cell 94:739–750

Yuan J, Horvitz HR (1990) The *Caenorhabditis elegans* gene *ced-3* and *ced-4* act cell autonomously to cause programmed cell death. Dev Biol 138:33–41

Yuan J, Horvitz HR (1992) The *Caenorhabditis elegans* cell death gene *ced-4* encodes a novel protein and is expressed during the period of extensive programmed cell death. Development 116:309–320

Yuan J, Shaham S, Ledoux S, Ellis HM, Horvitz HR (1993) The *C. elegans* cell death gene *ced-3* encodes a protein similar to mammalian interleukin-1β-converting enzyme. Cell 75:641–652

Zhou H, Henzel WI, Liu X, Luschg A, Wang X (1997) Apaf-1, a human protein homologous to *C. elegans* CED-4, participates in the cytochrome c-dependent activation of caspase-3. Cell 90:405–413

4 Presenilin Proteins and Their Function During Embryonic Development and Alzheimer's Disease

C. Haass

4.1 Introduction

Aggregation and precipitation of peptides appear to play a major role in neurodegenerative diseases such as Alzheimer's disease (AD) (Teplow 1998), Parkinson's disease (Mezey et al. 1998), and Huntington's disease (Georgalis et al. 1998). In AD, the aggregating Amyloid β-peptide (Aβ) accumulates in highly insoluble senile plaques, which are the defining pathological symptoms of the disease (Selkoe 1996). Aβ is derived by proteolytic processing from the β-Amyloid precursor protein (βAPP; Selkoe 1996). Two secretases have been postulated, which either generate the N-terminus (β-secretase) or the C-terminus (γ-secretase) of Aβ (Haass and Selkoe 1993; Fig. 1). Aβ is produced under physiological conditions in cultured cells and is secreted into the media. In vivo, Aβ is detected in human plasma and cerebrospinal fluid (Haass and Selkoe 1993).

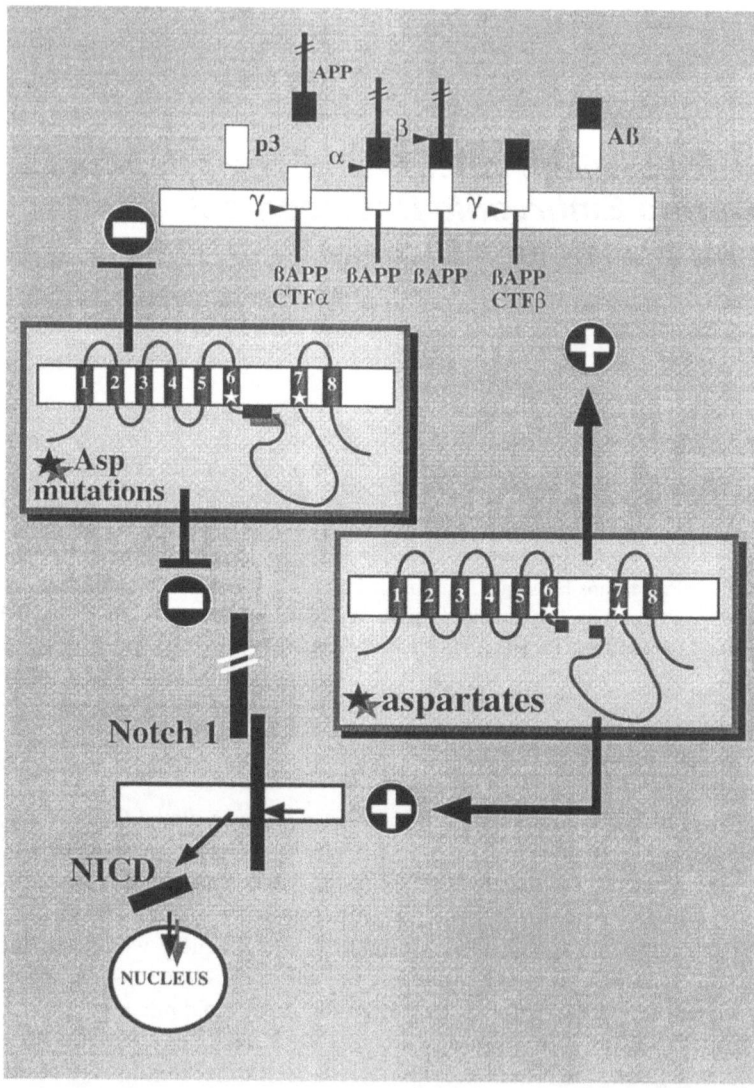

Fig. 1. Presenilin function in Notch signaling and proteolytic processing of βAPP (see text for details)

4.2 Familial Alzheimer's Disease

In the majority of cases AD occurs sporadically with an increasing risk during aging (Selkoe 1996; Price and Sisodia 1998). However, rare mutations have been found to cause autosomal-dominant early-onset familial AD (FAD; Selkoe 1996; Price and Sisodia 1998). Mutations were found within the genes encoding βAPP, presenilin-1 (PS), and PS2 (summarized in Selkoe 1996; Price and Sisodia 1998). The analysis of these mutations in primary fibroblasts derived from FAD patients, transfected cells, and transgenic animals revealed that they all alter the production of Aβ (Selkoe 1996; Price and Sisodia 1998). Interestingly, these mutations affect a common pathological mechanism by increasing the production of the long 42 amino acid version of Aβ (Aβ42). This peptide is known to aggregate much faster as compared to the more abundant Aβ40 (Jarret and Lansbury 1993) and is therefore predominantly found within senile plaques (Lemere et al. 1996; Mann et al. 1996).

4.3 The Presenilin Complex

PS proteins are membrane proteins, which most likely contain eight transmembrane domains (Doan et al. 1996; De Strooper 1997). Human presenilins are proteolytically processed to an N-terminal ~30 kDa fragment (NTF) and a C-terminal ~20 kDa fragment (CTF; Thinakaran et al. 1996; Fig. 1). In vivo, these fragments accumulate and almost no full-length PS can be observed (Thinakaran et al. 1996). PS fragments are bound to each other (Capell et al. 1998; Thinakaran et al. 1998) and form a high molecular weight complex (Capell et al. 1998). This complex, consisting of NTF and CTF, and probably other binding proteins as well (Yu et al. 1998; Zhang et al. 1998), may be the biologically active unit. Recent evidence indicates that recombinant NTF containing FAD-associated mutations does not stimulate Aβ42 production (Steiner et al. 1998; Citron et al. 1998; Tomita et al. 1998). Furthermore, such recombinant NTFs are also inactive in facilitating Notch signaling (Baumeister et al. 1997; see below) and coimmunoprecipitation experiments demonstrated that these fragments are not incorporated into the PS complex (Steiner et al. 1998). This raises the possibility that the PS complex formation composed of NTF and CTF may be required for the activity of PS proteins, a hypothesis which is strongly supported by the recently identified dominant negative mutations described below.

4.4 Presenilins Facilitate Aβ Production

Besides the pathological function of mutant PS in Aβ42 generation, wild-type PS proteins appear to have a fundamental function in physiological Aβ production, since it was recently found that PS1$^{-/-}$ mice produce significantly reduced amounts of Aβ40 and Aβ42 (De Strooper et al. 1998). Since C-terminal fragments of βAPP accumulate in neurons derived from the brains of PS$^{-/-}$ mice, it has been proposed that PS proteins may either activate γ-secretase (De Strooper et al. 1998; Haass and Selkoe 1998) or may even exhibit a γ-secretase activity by themselves. This proposal is based on the identification of two unusual intramembranous aspartate residues within transmembrane domains 6 and 7 (Wolfe et al. 1999; Fig. 1). Strikingly, when these aspartates were mutagenized the resulting PS variants completely blocked endoproteolysis of PS1. This was accompanied by a pronounced decrease in both major Aβ species, Aβ$_{40}$ and Aβ$_{42}$. Moreover, C-terminal fragments of βAPP produced either by α- or β-secretase accumulated to high levels. This indicates a major defect in the γ-secretase activity that would normally process these fragments to Aβ and p3, and hence the aspartate mutants have a dominant negative effect on γ-secretase activity. These findings are strikingly similar to the effects of a PS1 knockout in mice described above. The critical aspartate residues of PS1 are functionally conserved in human PS2 (Steiner and Haass, submitted) and zebrafish PS1 (Leimer and Haass, submitted). Based on these recent findings, Selkoe and colleagues presented a very exciting model in which PS proteins are unusual aspartyl proteases, and are in fact the long-sought γ-secretase (Fig. 1). Aspartyl proteases require two aspartate residues within their enzymatically active domain. Mutagenizing either one of them results in a complete loss of proteolytic activity. Therefore, presenilins may be autocatalytically activated, cleaving βAPP within the membrane to generate Aβ and p3 (Fig. 1).

4.5 Presenilins Are Required for Notch Signaling

PS proteins facilitate Notch signaling. This is supported by the finding that a defect in the PS homologue of *Caenorhabditis elegans* results in a Notch phenotype, which can be fully rescued by transgenic expression of human PS1 and PS2 (Baumeister et al. 1997; Levitan et al. 1996). Furthermore, a deletion of the PS1 gene in mice results in abnormal embryonic development causing a fatal phenotype, which is highly similar to that caused by the deletion of the Notch gene (Shen et al. 1997; Wong et al. 1997). Moreover, it has been recently demonstrated that cells derived from PS$^{-/-}$ mice fail to undergo the critical intramembranous cleavage of Notch (De Strooper et al. 1999; Fig. 1). This cleavage is required for the release of the Notch intracellular domain (NICD), which is then transported to the nucleus where it regulates gene transcription (Struhl and Adachi 1998; Schroeter et al. 1998). Interestingly, the dominant negative aspartate mutations described above also abolish the proteolytic release of NICD and the mutant cDNAs did not rescue the Notch phenotype in *C. elegans* with a defective endogenous PS gene (Steiner and Haass, submitted).

4.6 Presenilins: Targets for Anti-Alzheimer's Drugs?

Based on the results described above, it appears that PS proteins are involved in the intramembranous cleavage of a variety of membrane proteins such as βAPP, Notch, and APLP (which is a βAPP homologous protein), and it is very tempting to speculate that PS proteins are identical with the γ-secretase. Although this model is quite convincing, there are still many open questions, which may imply that PS molecules are involved in targeting selected membrane proteins to the cell surface (Naruse et al. 1998) and therefore affect βAPP and Notch processing rather indirectly. If indeed the PS proteins are identical with the γ-secretase, inhibition of PS function may be a major target for Amyloid-lowering drugs. However, due to the essential function of PS proteins in Notch signaling, inhibition of their biological activity may cause fatal side effects.

References

Baumeister R, Leimer U, Zweckbronner I, Jakubek C, Grünberg J, Haass C
    (1997) Human presenilin-1, nut no familial Alzheimer's disease (FAD mu-
    tants), facilitate Caenorhabditis elegans Notch signalling independently of
    proteolytic processing. Genes Funct 1:149–159
Capell A, Grünberg J, Pesold B, Diehlmann A, Citron M, Nixon R, Beyreuther
    K, Selkoe DJ, Haass C (1998) The proteolytic fragments of the Alzheimer's
    disease-associated presinilin-1 form heterodimers and occur as a 100-150-
    kDa molecular mass complex. J Biol Chem 273:3205–3211
Citron M, Eckman CB, Diehl TS, Corcoran C, Ostazewski BL, Xia W, St.
    George-Hyslop P, Younkin SG, Selkoe DJ (1998) Additive effect of PS1
    and APP mutations on secretion of the 42-residue Amyloid β-protein are
    mediated by the PS1 holoprotein. Neurobiol Dis 5:107–116
De Strooper B, Beullens M, Contreras B, Levesque L, Craessaerts K, Cordell
    B, Moechars D, Bollen M, Fraser P, St. George-Hyslop P, van Leuven F
    (1997) Phosphorylation, subcellular localization, and membrane orientation
    of the Alzheimer's disease-associated presenilins. J Biol Chem
    272:3590–3598
De Strooper B, Saftig P, Craessaerts K, Vanderstichele H, Guhde G, Annaert
    W, Von Figura K, van Leuven F (1998) Deficiency of presenilin-1 inhibits
    the normal cleavage of amyloid precursor protein. Nature 391:387–390
DeStrooper B, Annaert W, Cupers P, Saftig P, Craessaerts K, Mumm J,
    Schroeter E, Schrijvers V, Wolfe M, Ray W, Goate A, Kopan R (1999) A
    presenilin-1-dependent γ-secretase-like protease mediates release of Notch
    intracellular domain. Nature 398:518–522.
Doan A, Thinakaran G, Borchelt DR, Slunt HH, Ratovitsky T, Podlisny M,
    Selkoe DJ, Seeger M, Gandy SE, Price DL, Sisodia SS (1996). Protein to-
    pology of presenilin 1. Neuron 17:1023–1030
Georgalis Y, Starikov EB, Lurz R, Scherzinger E, Saenger W, Lehrach H,
    Wanker EE (1998) Huntingtin aggregation monitored by dynamic light
    scattering. Proc Natl Acad Sci USA 95:6118–6121
Haass C, Selkoe DJ (1993) Cellular processing of beta-amyloid precursor pro-
    tein and the genesis of amyloid beta-peptide. Cell 75:1039–1042
Haass C, Selkoe DJ (1998) Alzheimer's disease. A technical KO of amyloid-
    beta peptide. Nature 391:339–340
Jarret JT, Lansbury PT Jr (1993) Seeding "one-dimensional chrystallization"
    of amyloid: a pathogenic mechanism in Alzheimer's disease and scrapie?
    Cell 73:1055–1058
Lemere CA, Lopera F, Kosik KS, Lendon CL, Ossa J, Saido TC, Yamaguchi
    H, Ruiz A, Martinez A, Madrigal L, Hincapie L, Arango JC, Koo EH, Goate
    AM, Selkoe DJ, Arango JC (1996) The E280A presenilin 1 Alzheimer mu-

tation produces increased A beta 42 deposition and severe cerebellar pathology. Nat Med 2:1146–1150

Levitan D, Doyle T, Brousseau D, Lee M, Thinakaran G, Slunt H, Sisodia S, Greenwald I (1996) Assessment of normal and mutant human presenilin function in Caenorhabditis elegans. Proc Natl Acad Sci USA 93:14940–14944

Mann DM, Iwatsubo T, Cairns NJ, Lantos PL, Nochlin D, Sumi SM, Bird TD, Poorkaj P, Hardy J, Hutton M, Prihar G, Crook R, Rossor MN, Haltia M (1996) Amyloid beta protein (Abeta) deposition in chromosome 14-linked Alzheimer's disease: predominance of Abeta42(43). Ann Neurol 40:149–156

Mezey E, Dehejia A, Harta G, Papp MI, Polymeropoulos P, Brownstein MJ (1998) Alpha synuclein in neurogenerative disorders: Murderer or accomplice? Nat Med 4:755–757

Naruse S, Thinakaran G, Luo J-J, Kusiak JW, Tomita T, Iwatsubo T, Qian X, Ginty DD, Price DL, Borchelt DR, Wong, PC, Sisodia SS (1998) Effects of PS1 deficiency on membrane protein trafficking in neurons. Neuron 21:1213–1221

Price D, Sisodia S (1998) Mutant genes in familial Alzheimer's disease and transgenic models. Annu Rev Neurosci 21:479–505

Schroeter EH, Kisslinger JA, Kopan R. (1998) Notch-1 signalling requires ligand-induced proteolytic release of intracellular domain. Nature 393:382–386

Selkoe DJ (1996) Amyloid beta-protein and the genetcs of Alzheimer's disease. J Biol Chem 271:18295–18298

Shen J, Bronson RT, Chen DF, Xia W, Selkoe DJ, Tonegawa S (1997) Skeletal and CNS defects in Presenilin-1-deficient mice. Cell 89:629–639

Steiner H, Capell A, Pesold B, Citron M, Kloetzel P-M, Selkoe DJ, Romig H, Mendla K, Haass C (1998) Expression of Alzheimer's diesease-associated presenilin-1 is controlled by proteolytic degradation and complex formation. J Biol Chem 273:32322–32331

Struhl G, Adachi A (1998) Nuclear access and action of notch in vivo. Cell 93:649–660

Teplow D (1998) Structural and kinetic features of amyloid β-protein fibrillogenesis. Amyloid 5:121–142

Thinakaran G, Borchelt DR, Lee MK, Slunt HH, Spitzer L, Kim G, Ratovitsky T, Davenport F, Nordstedt C, Seeger M, Hardy J, Levey AI, Gandy SE, Jenkins NA, Copeland NG, Price DL, Sisodia SS (1996) Endoproteolysis of presenilin 1 and accumulation of processed derivates in vivo. Neuron 17:181–190

Thinakaran G, Regard JB, Bouton CM, Harris CL, Price DL, Borchelt DR, Sisodia SS (1998) Stable association of presenilin derivates and absence of presenilin interactions with APP. Neurobiol Dis 4:438–453.

Tomita T, Tokuhiro S, Hashimoto T, Aiba K, Saido T, Maruyama K, Iwatsubo T (1998) Molecular dissection of domains in mutant presenilin 2 that mediate overproduction of amyloidogenic forms of amyloid beta peptides. Inability of trunctated forms of PS2 with familial Alzheimer's disease mutation to increase secretion of Abeta42. J Biol Chem 273:21153–21160

Wolfe M, Xia W, Ostaszewski B, Diehl T, Kimberly W, Selkoe D (1999) Two transmembrane aspartates in presenilin-1 required for presenilin endoproteolysis and γ-secretase activity. Nature 398:513–517.

Wong PC, Zheng H, Chen H, Becher MW, Sirinathsinghji DJS, Trumbauer ME, Chen HY, Price DL, Van der Ploeg LHT, Sisodia SS (1997) Presenilin 1 is required for Notch 1 and DII1 expression in the paraxial mesoderm. Nature 387:288–292

Yu G, Chen F, Levesque G, Nishimura M, Zhang DM, Levesque L, Rogaeva EA, Xu M, Liang Y, Duthie M, St. George-Hyslop PH, Fraser PE (1998) The presenilin 1 protein is a comonent of a high molecular weight intracellular complex that contains beta-catenin. J Biol Chem 273:16740–16475

Zhang Z, Hartmann H, Do VM, Abramowski D, Strchler-Pierrat C, Staufenbiel M, Sommer B, van de Wetering M, Clevers H, Saftig P, De Strooper B, Yankner BA (1998) Destabilization of beta-catenin by mutations in presenilin-1 potentiates neuronal apoptosis. Nature 395:698–702

5 Cell–Cell Interaction During Drosophila *Embryogenesis*: *Novel Mechanisms and Molecules*

M. Affolter

5.1 Cell–Cell Interactions in Development

During the development of multicellular organisms, cells interact extensively in order to generate tissues and organs at a defined time in the appropriate place. In the past 20 years, numerous cell–cell signaling systems have been identified and characterized. It has become a general rule that a given signaling system is used repetitively during development, in the same, as well as in different, cellular contexts. However, the cellular response to the signal varies in most cases and appears to rely upon the developmental history of the responding cell.

To better understand cell–cell interactions as they occur in developing multicellular systems, we have focused our studies on the molecular characterization of genes involved in the formation of two tissues: the endodermal cell layer and the developing tracheal system. The patterning of the endodermal cell layer depends on induction between two single-celled sheets and represents a relatively simple situation. In contrast, the development of the tracheal system via cell migration and cell shape changes, depends on numerous interactions between different groups of cells and the surrounding extracellular substrates. In both systems, a member of the TGFβ superfamily of secreted signaling molecules plays an important role which will be highlighted and will serve as a guide through this chapter.

5.1.1 Induction in Development:
A Genetic Approach in *Drosophila melanogaster*

One of the major challenges of embryological studies over the last decades has been the determination of the molecular mechanisms by which the spatial organization of an animal emerges as it develops from a fertilized egg. A leading role in this process has been attributed to mechanisms by which a signal generated by a group of cells controls the fate of a neighboring cell(s), a phenomenon generally referred to as induction. In vertebrates, where induction was first described, most organs are formed through interactions between cells of different germ layers. In invertebrates, genetic studies have demonstrated that induction is also one of the prevailing mechanisms that steers developmental decisions.

Using the powerful genetic model of *Drosophila melanogaster*, we have been studying embryonic induction processes occurring between cells of two different cell layers (Bienz 1994, 1996). As a result of these inductions, positional information laid down in the visceral mesoderm (VM) is transmitted via cell–cell communication to the underlying endoderm, leading to its patterning along the anterior-posterior axis. A particular induction cascade culminating in the development of copper cells in the responding endoderm is mediated in part by DECAPENTA-PLEGIC (DPP), a member of the TGFβ superfamily of secreted signaling molecules. Based on studies in vertebrates by the group of Joan Massagué, we have isolated, in collaboration with Konrad Basler's group in Zürich, the two essential cell surface receptors which bind the DPP ligand and transmit the signal across the membrane (Nellen et al. 1994; Ruberte et al. 1995). These receptors were encoded by the genes PUNT (PUT, a type II receptor) and THICK VEINS (TKV, a type I receptor). Mutations in these genes were previously identified by Nüsslein-Volhard and colleagues in their screens for zygotic effect embryonic lethal mutations. Helped by mutant phenotypes observed in the absence of these essential DPP receptors, we subsequently identified a novel nuclear component (SCHNURRI), which is essential in most cells for the generation of an appropriate nuclear response to DPP (Grieder et al. 1995). The *schnurri* gene encodes a long protein of more than 2500 amino acids containing eight zinc fingers; it presumably assists other factors in the generation of appropriate transcriptional changes induced by DPP in responding cells. Although we have invested a considerable effort in the genetic and biochemical analysis of *schnurri*, we do not know with certainty whether SCHNURRI binds directly to *cis*-acting elements of DPP-responsive genes or plays a more indirect role in DPP signaling.

At the same time as we and others reported the identification of the cell surface receptors for DPP, Bill Gelbart's group (Harvard) reported the identification of two genes, *mad* and *medea*, which when mutated, modify the response of cells to DPP (Raftery et al. 1995; Sekelsky et al. 1995). Subsequently, the biochemical characterization of the vertebrate homologues of MAD (invertebrate and vertebrate proteins are now collectively referred to as SMADS) showed that they encode a class of proteins that function as intracellular signaling effectors for the TGFβ superfamily of secreted polypeptides.

The two essential DPP receptors PUT and TKV, and the signal transducers MAD and MEDEA appear to constitute the core components of the DPP signaling pathway and are required in all cells to transmit the DPP signal. How is it possible then that cells respond differently to the reception of the DPP molecule if they use the same signal transduction components? The transcriptional responses to DPP in the developing midgut are extremely diverse and depend on the germ layer provenance of a given cell as well as on the parasegmental origin of cells within the visceral mesoderm. For example, visceral mesoderm cells of parasegment 7 induce the transcription of the homeotic gene *Ubx* and the *dpp* gene itself as a response to DPP signaling, parasegment 8 cells induce *wg* transcription, whereas parasegment 6 cells induce neither of these genes. Endodermal cells induce expression of the homeotic gene *labial* (*lab*) upon stimulation with DPP.

5.1.2 HOX Genes and Positional Signaling: Is There a Link?

One transcriptional target of DPP-mediated endoderm induction is the homeotic gene *lab*. To investigate the potential role of SCHNURRI in regulating *lab* activation, we generated a large number of mutant versions of the DPP-dependent enhancer of *lab*, called lab550. Although this analysis has not yet allowed us to draw any firm conclusions concerning the function of SCHNURRI, we made a most interesting and intriguing observation. We found that the activity of the enhancer is stimulated significantly only by DPP signaling upon binding of the LAB homeotic protein and its cofactor EXTRADENTICLE (EXD) to an essential and evolutionarily conserved sequence element within the 5' part of the enhancer (Grieder et al. 1997). Thus, the lab550 enhancer appears to integrate positional information (via the region-specific expression of the homeotic gene *lab*) and spatiotemporal information (via DPP signaling); only when these inputs act in concert in an endodermal cell is the enhancer fully active. These results illustrate how synergistic effects on an enhancer carrying both DPP- and HOM-responsive sequences can contribute to the generation of a specific cellular response to DPP; only cells expressing *lab* can activate this particular DPP-responsive enhancer efficiently.

When these results are interpreted with respect to the possible function of homeotic genes during development, it emerges that one of the very important roles of homeotic proteins (which are the classic example of proteins conveying positional information to cells in a developing embryo) is to participate in the interpretation of signaling inputs into appropriate, position-specific transcriptional responses. Although regulatory cross-talk is a common occurrence in the control of gene transcription, there is little or no precedent for the interaction of homeotic proteins with those involved in signaling, and I am not aware that such a direct "signal interpretation" function of the homeotic proteins has been proposed previously.

Such a proposal is very interesting for discussion in the context of the developing midgut. As already mentioned, DPP not only induces the expression of specific genes in the endodermal cell layer in the *Drosophila* embryo, but also in the VM cell layer itself. Genetic studies provided evidence that the cellular transcriptional response to DPP in the VM largely depends on the parasegmental provenance of, and thus on the expression of, a specific homeotic gene in a given cell (Bienz 1994, 1996). For example, the DPP-induced transcription of *wingless* (*wg*) in anterior cells of parasegment 8 (which border the DPP-secreting parasegment 7 cells) requires the activity of the homeotic gene *abdA*. In analogy to the lab550 enhancer, it is likely that the DPP-dependent enhancer of the *wg* gene requires a direct interaction with ABDA. Homeotic proteins and signaling are also tightly linked to the development of segment-specific epidermal structures. The development of the leg imaginal discs is repressed in abdominal segments by the homeotic genes of the bithorax complex (Vachon et al. 1992). The signals involved in the generation of these disc primordia are also secreted and transduced in the abdominal segments, but the expression of the homeotic proteins of the bithorax complex changes the way cells interpret these signaling inputs in the posterior segments. A similar situation occurs in the developing larva. Although the global coordinate systems of the wing and the haltere are the same, UBX appears to interpret signaling input differently in the haltere (Weatherbee et al. 1998; Halder et al. 1998). Recently, genetic studies in *C. elegans* showed that the Hox gene *lin-39* is required during vulval induction to select the outcome of RAS signaling (Maloof and Kenyon 1998). Therefore, the transcriptional targets of signaling pathways might be selectively activated or repressed with the help of the homeotic protein present in a given cell.

Based on our findings and these general considerations, a major focus of our future efforts will be centered around the elucidation of the molecular mechanisms underlying the synergistic activation of the *lab* enhancer by DPP signaling and by HOM/EXD, or, put more generally, the synergy between signaling and HOX input. We will continue to dissect the enhancer in vivo, and use biochemical methods as well as in vivo systems to analyze possible molecular interactions between molecules of the DPP signaling pathway and homeotic proteins or their cofactors. Experiments are underway which should allow us to determine the site(s) on lab550 through which the DPP signaling cascade acts. In addition, we will try to identify the factor(s) which contribute to the observed tissue specificity of the lab550 enhancer. We anticipate to learn how specific nuclear responses to DPP are generated, what factors are involved (signaling components, homeotic proteins, germ layer specific factors etc.), and how these factors interact on DPP-response elements. Since homeotic genes are region-specifically expressed in many organs which are built as a result of local induction events, our studies might provide insight into the basic mechanisms by which induction shapes tissues and organs in vertebrates.

A few years ago, we became interested in the development of the tracheal system, a network of oxygen-transporting tubules, during *Drosophila* embryogenesis. Tracheal development relies on cell migration and cell extension (in the absence of concomitant cell division), and we anticipated that the establishment of this complex tubular system would require numerous cell–cell and cell–substrate interactions. In the meantime, tracheal development has emerged as a valuable model system to study the genetics of cell migration in *Drosophila*. The proposition that complex cellular interactions would help in establishing the tubular network have been confirmed, and I would like to concentrate in the second part of this chapter on the results we and other groups have obtained regarding the genetic control of the establishment of branched tubular epithelia in *Drosophila*.

5.2 Cell Migration in Development

During the development of multicellular animals, a large number of cells originate at a considerable distance from the site within the organism at which they fulfill their essential function. To reach that site, these cells have to migrate over other tissues or extracellular substrates receiving and interpreting instructions from their environment. Such precise cell movements are a prerequisite for the concerted development and the subsequent interconnection of tissues within developing organisms. Although a coherent picture of how cells generate force and motility is beginning to emerge (Lauffenburger and Horvitz 1996; Mitchison and Cramer 1996), the understanding of how mobility is regulated during development is less advanced. Locally distributed signals dictating the direction of migration are most likely molecules anchored to cell membranes of the migration surface, molecules associated with the extracellular environment of the migration path, or diffusible factors. Receptors recognizing these guidance molecules as attractive or repulsive must be expressed on the surface of migrating cells and their activation should ultimately lead to the intracellular changes that cause directed migration.

Recently, molecules with cell migration guidance properties have been identified in invertebrates and vertebrates. Genetic studies in *C. elegans* have led to the identification of a gene (*unc-6*) which is required for mesoblast migration and pioneer axon extension in dorsal and ventral directions on the body wall (Hedgecock et al. 1990). *unc-6* encodes a secreted protein with an N-terminus homologous to laminin subunits and is now generally referred to as a netrin. Multiple netrin cues are important for proper regulation of the dorsal guidance receptor UNC-5, a cell surface receptor of the immunoglobulin superfamily which mediates some cellular responses to the UNC-6 guidance cues (Hamelin et al. 1993). More recently, netrins have been identified in vertebrates and in *Drosophila* where they are involved in axon guidance (Hedgecock and Norris 1997). The netrins identified thus far in *Drosophila* were not involved in directed cell migration during tracheal development (unpublished results). Orthologs for several other vertebrate guidance molecules or their receptors (Wehrle-Haller and Weston 1997) have not been found yet in *Drosophila,* therefore it has not been possible to test their role in tracheal development.

5.2.1 Control of Cell Migration During Tracheal Development in *Drosophila*

Due to our interest in intercellular signaling, we started to study trachea formation during embryonic development in *Drosophila melanogaster* a few years ago, with special emphasis on the genetic control of directed cell migration. The stereotyped tubular network of the larval trachea develops from ten individual small bulges containing approximately 90 cells, the tracheal placodes, which form in lateral positions in the trunk segments on either side of the embryo (Manning and Krasnow 1993; Samakovlis et al. 1996). Upon invagination of the tracheal cells (generating a tracheal sac), the complex branching pattern of the tracheal system is established via cell migration, extension, and fusion in the absence of further cell division. During this process, a defined number of cells migrate from each placode as anterior branches and as ventral and dorsal branches (e.g., five to seven cells migrate dorsally and eventually line up to form the dorsal branch which targets the dorsal vessel and the dorsal muscles). Thus, the establishment of the tracheal network requires that groups of cells are instructed with respect to distinct routes, which they have to follow. The highly stereotyped development of the tracheal system combined with the powerful genetic model of *Drosophila* should allow a molecular dissection of this process and lead to a better understanding of how cells recognize and interpret migration cues. One of our main interests is to find out whether migration during tracheal development is controlled in an axis-specific fashion, as has been observed for a number of cells in *C. elegans* (see above).

Mutations in a number of genes have been reported to cause abnormal development of the tracheal system. The *breathless* (*btl*) gene is expressed in all tracheal cells and encodes a *Drosophila* fibroblast growth factor receptor (DFGF-R; Klämbt et al. 1992). In the absence of *btl* function, tracheal cell division as well as general tracheal cell fate is unaffected, but the cells fail to migrate and consequently no branches are formed. Initial studies with activated components of the FGF signaling pathway suggested that the normal activity of *btl* in promoting cell migration does not require spatially restricted cues (Reichmann-Fried et al. 1994). However, more recent results indicate that BTL is activated by a spatially restricted FGF-like ligand, the product of the *branchless* (*bnl*) gene (Sutherland et al. 1996). The dynamic expression of *bnl* at each position where a new branch will form and grow out, combined with the results obtained from ectopic expression experiments, suggest that BNL acts as a chemoattractive guidance molecule (Sutherland et al. 1996).

5.2.2 Isolation of Genes
Required for Directed Tracheal Cell Migration

To isolate components required for directed cell migration, we have analyzed, over the last five years, a large number of mutations for possible tracheal defects using as a tool an antibody that outlines the developing trachea. In these various screens, we have isolated mutations giving rise to two qualitatively different tracheal phenotypes: (a) tracheal cell migration completely absent, and (b) tracheal cell migration defective in distinct directions only. The characterization of these mutations and the isolation of the corresponding wild-type gene products will be discussed separately.

5.2.3 Isolation of a Novel Component Required for FGF Signaling

We have isolated three mutants in which tracheal cells are normally specified but completely fail to migrate. The mutation in one strain occurred in the *btl* gene; a second mutant strain carried a deletion which removed the *bnl* locus (unpublished observations). The mutation in the third strain was located genetically with the aid of a deficiency "kit" to chromosomal region 88 on the third chromosome. In collaboration with the group of Maria Leptin (Cologne), we have recently identified the gene corresponding to the mutated locus in this strain and shown that it encodes a novel protein containing two ankyrin repeats and a coiled-coil region (Vincent et al. 1998). The gene is transcribed specifically in two tissues, the trachea and the mesoderm; in both tissues, expression starts very early, during or just after cell determination. Interestingly, both tissues rearrange through extensive cell migrations requiring FGF signaling (through the FGF receptors *btl* in the trachea and *heartless (htl)* in the mesoderm; Beiman et al. 1996; Gisselbrecht et al. 1996). Inspection of the mesodermal phenotype in embryos of our mutant showed that in addition to tracheal cells, mesodermal cells completely failed to migrate. Using a large number of antibodies as markers, we found that both in mesodermal and tracheal cells of the mutants, the FGF receptors *htl* and *btl,* respectively, fail to activate the RAS-MAP kinase pathway (through which the FGF receptor signal is propagated; Reichman-Fried et al. 1994). Genetic epistasis experiments demonstrated that the novel gene acts downstream of the activated receptor but upstream of *ras.* Based on these findings we named the gene *Downstream of FGFR (dof).* Our studies strongly suggest that the novel DOF protein links the activated FGF receptors to the RAS/MAPK signaling cascade and possibly to other cytoplasmic components. A protein with a similar function, FRS2/SNT, has recently been identified in vertebrate cells (Kouhara et al. 1997); its genetic requirement for FGF signaling remains to be demonstrated. The possible function of DOF will be discussed further at the end of the chapter in the context of all other components which have been shown to be required for directed migration.

5.2.4 Identification of Genes Required for the Migration of Specific Tracheal Branches

As mentioned above, we have also isolated a number of mutants which, in contrast to *bnl, btl,* and *dof,* lack specific branches only. These mutations appeared very interesting to us since they disturb migration along distinct directions, their effects being comparable to those observed in *unc-5* and *unc-6* mutants in *C. elegans*, in which migrations along the dorsal-ventral axis are selectively disturbed (see references in Wadsworth et al. 1996). It turned out that two of the mutations which affected the development of branches growing dorsally and ventrally, exclusively, mapped to the genes encoding the two essential DPP receptors TKV and PUT (Affolter et al. 1994; Ruberte et al. 1995). Our detailed analysis of tracheal development in these two mutants (using many tools generated in the course of our studies on midgut development) revealed that DPP plays a dual role during tracheal cell migration (Vincent et al. 1997; see also Wappner et al. 1997; Llimargas et al. 1997). On the one hand, DPP controls the region-specific activation of *bnl* in the dorsal part of the embryo. On the other hand, DPP expression dorsal and ventral to the tracheal placode at the onset of migration, instructs small groups of tracheal cells with respect to their migration behavior. In the absence of DPP signaling, dorsal and ventral tracheal cells lose their ability to migrate along the dorsal-ventral body axis. Ectopic DPP signaling in the center of the tracheal placode inhibits anterior migration, and reprograms cells to adopt a dorsal-ventral migration behavior. Our studies suggest that other factors in addition to BNL dictate the direction of migration along the dorsal-ventral axis; some of these factors might be recognized by tracheal cells only upon the reception of the DPP signal, and might be distributed in an axis-specific fashion during early embryogenesis.

Further studies in collaboration with the laboratory of Reinhard Schuh in Göttingen demonstrated that two genes encoding the zinc finger transcription factors KNIRPS and KNIRPS RELATED possess multiple and redundant functions during tracheal development. *knirps/knirps related* expression is induced in dorsal and ventral tracheal cells as a result of DPP receptor activation. KNIRPS and KNIRPS RELATED are required for the migration of cells forming the dorsal and ventral branches. In addition, ectopic *knirps* or *knirps related* expression in lateral tracheal cells respecifies their anteroposterior migration to a dorsoventral migration behavior (Chen et al. 1998).

5.2.5 Guiding Migration of Tracheal Cells: A Working Model

The genetic and molecular studies of embryonic tracheal development have provided a relatively detailed picture of how the major tracheal branches are established after the invagination of the tracheal placode. Clearly, one of the major determinants of the primary branching pattern is the FGF ligand encoded by the *bnl* gene. The dynamic spatial expression of BNL in ectodermal and mesodermal cells around the tracheal sac prefigures the direction of migration of tracheal cells. Gain-of-function experiments are entirely consistent with a chemoattractive function of BNL. But how does the spatial activation of the BNL receptor BTL, a transmembrane receptor tyrosine kinase, lead to the directed outgrowth of tracheal cells? One important novel and essential component of BTL- (and HTL-)mediated migration is encoded by the *dof* gene. DOF links the activated receptor to the RAS-MAPK cascade but it is unlikely that this represents the only function of DOF. DOF might also act as a mediator between the activated receptor and cytoskeletal components, ultimately leading to a polarization of the responding cells. Based on experiments with activated hybrid receptors, it has indeed been argued that the activation of the RAS-MAPK cascade is not sufficient for directed tracheal migration (Lee et al. 1997). It is likely that local BTL signaling polarizes the responding cells, leading to the local activation of a cellular locomotion "machinery" (Mitchison and Cramer 1996; Lauffenburger and Horvitz 1996). The isolation and characterization of DOF provides a promising entry point to study how the FGFRs BTL and BNL affect cellular migration behaviors. At present, we are trying to isolate proteins that interact with DOF in an attempt to link receptor activation to directed migration.

Although it remains to be shown whether FGFR signaling to the nucleus is important for directed migration, it is clear that DPP signaling in dorsal and ventral cells is essential for their directed migration and is mediated via the transcriptional induction of two genes which in turn encode transcription factors (KNIRPS and KNIRPS RELATED; Chen et al. 1998). Dorsal and ventral tracheal cells have to receive and respond to the DPP signal in order to be attracted and migrate towards the dorsal and ventral spots of BNL. Similarly, cells migrating anteriorly to form the dorsal trunk have to respond to EGF signaling which leads to the activation/maintenance of expression of the transcription factor SPALT; SPALT enables dorsal trunk cells to move towards the anteriorly located BNL spot (Kühnlein et al. 1996; Wappner et al. 1997; Chen et al. 1998).

The genes regulated by KNIRPS/KNIRPS RELATED and SPALT remain to be identified. These targets might encode components which are essential for the proper interpretation of FGFR signaling, components which regulate the affinity of groups of tracheal cells, or components which recognize essential migration cues distinct from BNL. We have already identified several transcription units that are under the control of KNIRPS and are in the process of cloning and characterizing the corresponding genes.

The developing tracheal system represents a unique system to study cell migration in *Drosophila* since both a guidance molecule (BNL) and its receptor (BTL) have been identified. Several other components (DOF, DPP signaling, EGF signaling, KNIRPS/KNIRPS RELATED, and SPALT) have been isolated and a future challenge will be to integrate all these components into a comprehensive network of interactions controlling directed cell migration.

5.3 Prospects

Many similarities between the development of the tracheal system of *Drosophila* and the vertebrate vascular system are emerging. Particularly striking is the interaction between TGFβ and FGF in cell migration during angiogenesis (Gadjusek et al. 1993), which argues for a more general role of the combined activity of these two signaling pathways in inducing cell migration. Even more striking is the conservation of aspects of the molecular control of branching in the *Drosophila* tracheal system and vertebrate lung development, and the involvement of FGF in the initial dispersion of mesodermal cells during early mouse development. It will be interesting to compare the regulatory networks upstream and downstream of FGFs in the distinct developmental contexts. These studies might ultimately lead to a better understanding of basic molecular strategies in development and disease.

References

Affolter M, Nellen D, Nussbaumer U, Basler K (1994) Multiple requirements for the receptor serine/threonine kinase *thick veins* reveal novel functions of TGFβ homologs during *Drosophila* embryogenesis. Development 120:3105–3117

Beiman M, Shilo BZ, Volk T (1996) Heartless, a *Drosophila* FGF receptor homolog, is essential for cell migration and establishment of several mesodermal lineages. Genes Dev 10:2993–3002

Bienz M (1994) Homeotic genes and positional signalling in the *Drosophila* viscera. Trends Genet 10:22–26

Bienz M (1996) Induction of the endoderm in *Drosophila*. Sem Cell Dev Biol 7:113–119

Chen C-K , Kühnlein RP, Eulenberg KG, Vincent S, Affolter M, Schuh R(1998) The transcription factors KNIRPS and KNIRPS RELATED control cell migration and branch morphogenesis during *Drosophila* tracheal development. Development 125:4959–4968

Gadjusek, CM, Luo Z, Mayberg MR (1993) Basic fibroblast growth factor and transforming growth factor beta-1: synergistic mediator of angiogenesis in vitro. J Cell Physiol 157:133–144

Gisselbrecht S, Skeath JB, Doe CQ, Michelson AM (1996) *heartless* encodes a fibroblast growth factor receptor (DFR1/DFGF-R2) involved in the directional migration of early mesodermal cells in the *Drosophila* embryo. Genes Dev 10:3003–3017

Grieder NC, Nellen D, Burke R, Basler K, Affolter M (1995) *schnurri* is required for *Drosophila* Dpp signaling and encodes a zinc finger protein similar to the mammalian transcription factor PRDII-BF1. Cell 81:791–800

Halder G, Polaczky P, Kraus ME, Hudson A, Kim J, Laughon A, Carroll S (1998) The vestigal and scalloped proteins act together to directly regulate wing-specific gene expression in *Drosophila*. Genes Dev 12:3900–3909

Hamelin, Mzhou Y, Su MW, Culotti, JG (1993) Expression of the UNC-5 guidance receptor in the touch neurons of C. elegans steers their axons dorsally. Nature 364:327–330

Klämbt C, Glazer L, Shilo BZ (1992) *breathless*, a *Drosophila* FGF receptor homolog, is essential for migration of tracheal and specific midline glial cells. Genes Dev 6:1668–1678

Kouhara H, Hadari YR, Spivak-Kroizman T, Schilling J, Bar-Sagi D, Lax I, Schlessinger J (1997) A lipid-anchored Grb2-binding protein that links FGF-receptor activation to the Ras/MAPK signaling pathway. Cell 89:693–702

Kühnlein RP, Schuh R (1996) Dual function of the region-specific homeotic gene *spalt* during *Drosophila* tracheal system development. Development 122:2215–2223

Lauffenburg DA, Horwitz AF (1996) Cell migration: A physically integrated molecular Process. Cell 84:359–369

Lee T, Hacohen N, Krasnow M, Montell DJ (1996) Regulated Breathless receptor tyrosine kinase activity required to pattern cell migration and branching in the *Drosophila* tracheal system. Genes Dev 10:2912–2921

Llimargas M, Casanova J (1997) *ventral veinless*, a POU domain transcription factor, regulates different transduction pathways required for tracheal branching in *Drosophila*. Development 124:3273–3281

Maloof J, Kenyon C (1998) The Hox gene *lin-39* is required during *C. elegans* vulval induction to select the outcome of Ras signaling. Development 125:181–190

Manning G, Krasnow MA (1993) Development of the *Drosophila* tracheal system. In: M. Bate and A. Martinez Arias (eds) The Development of *Drosophila melanogaster*. Cold Spring Harbor Laboratory Press, New York, pp 609–685

Michelson AM, Gisselbrecht S, Zhou Y, Baek KH, Buff EM (1998) Dual functions of the Heartless fibroblast growth factor receptor in development of the *Drosophila* embryonic mesoderm. Dev Genet 22:212–229

Mitchison TJ, Cramer LP (1996) Actin-based cell motility and cell locomotion. Cell 84:371–379

Raftery L, Twombly V, Wharton K, Gelbart W (1995) Genetic screens to identify elements of the *decapentaplegic* pathway in *Drosophila*. Genetics 139:241–254

Reichman-Fried M, Dickson B, Hafen E, Shilo BZ (1994) Elucidation of the role of breathless, a *Drosophila* FGF receptor homolog, in tracheal cell migration. Genes Dev 8:428–439

Ruberte E, Marty T, Nellen D, Affolter M, Basler K (1995) An absolute requirement for both the type II and type I receptors, punt and thick veins, for dpp signaling in vivo. Cell 80:890–898

Samakovlis C, Hacohen N, Manning G, Sutherland DC, Guillemin K, Krasnow MA (1996) Development of the *Drosophila* tracheal system occurs by a series of morphologically distinct but genetically coupled branching events. Development 122:1395–1407

Sekelsky JJ, Newfeld SJ, Raftery LA, Chartoff EH, Gelbart WM (1995) Genetic characterization and cloning of *mothers against dpp*, a gene required for decapentaplegic function in *Drosophila melanogaster*. Genetics 139:1347–1358

Shishido E, Ono N, Kojima T, Saigo K (1997) Requirements of DFR1/Heartless, a mesoderm-specific *Drosophila* FGF- receptor, for the formation of heart, visceral and somatic muscles, and ensheathing of longitudinal axon tracts in CNS. Development 124:2119–2128

Sutherland D, Samakovlis C, Krasnow MA (1996) *branchless* encodes a *Drosophila* FGF homolog that controls tracheal cell migration and the pattern of branching. Cell 87:1091–1101

Vachon G, Cohen B, Pfeifle C, McGuffin ME, Botas J, Cohen S (1992) Homeotic Genes of the Bithorax Complex Repress Limb Development in the Abdomen of the *Drosophila* Embryo through the Target Gene *Distal-less*. Cell 71:437–450

Vincent S, Ruberte E, Grieder NC, Chen CK, Haerry T, Schuh R, Affolter M (1997) DPP controls tracheal cell migration along the dorsoventral body axis of the *Drosophila* embryo. Development 124:2741–2750

Wappner P, Gabay L, Shilo BZ (1997) Interaction between the EGF receptor and DPP pathways establish distinct cell fates in the tracheal placodes. Development 124:4707–4716

Weatherbee S, Halder G, Kim J, Hudson A, Carroll S (1998) Ultrabithorax Regulates Genes at Several Levels of the Wing-Patterning Hierarchy to Shape the Development of the *Drosophila* Haltere. Genes Dev 10:1474–1482

Wehrle-Haller B, Weston JA (1997) Receptor tyrosine kinase-dependent neural crest migration in response to differentially localized growth factors. Bioessays 19:337–345

6 The Roles of BMPs, BMP Antagonists, and the BMP Signaling Transducers Smad1 and Smad5 During Dorsoventral Patterning of the Zebrafish Embryo

M. Hild, A. Dick, H. Bauer, S. Schulte-Merker, P. Haffter,
T. Bouwmeester, and M. Hammerschmidt

6.1 Early Zebrafish Development and Embryonic Patterning

Development of the zebrafish starts with a single cell, the fertilized egg, also called the zygote. After fertilization, the embryo undergoes a rapid succession of cell cleavages, cell specifications, and morphogenetic events which are strictly coordinated temporally and spatially, and which within two days lead to a larva that very much resembles the adult fish. Important morphogenetic processes occur during gastrulation when the embryo becomes multilayered (Warga and Kimmel 1990). Three germ layers – ectoderm, endoderm, and mesoderm – become evident, and endoderm and mesoderm are moved into the interior of the embryo. During further development, the ectoderm will give rise to skin and fins (epidermal fates) and to the central nervous system; the endoderm to the digestive system (stomach, gut, liver, pancreas etc.); and the mesoderm to muscle, the skeletal system, the blood, the vascular system, and the remaining inner organs such as heart and kidney (pronephros).

Before gastrulation, during blastula stages, the embryo is rather simply structured and appears radially symmetrical, with an animal-vegetal polarity, reflecting the presumptive anteroposterior axis, but with no morphologically apparent dorsoventral polarity. Despite this morphological symmetry, however, several genes display a differential dorsoventral expression pattern, indicating that positional values along the presumptive dorsoventral axis are already defined at pregastrula stages. In addition, cell lineage-tracing experiments indicate that fates of cells depend on their position within the early blastula embryo (Kimmel et al. 1990). Along the animal-vegetal axis, the fate map of the blastula embryo is rather simple; cells in animal regions of the blastula embryo give rise to ectodermal fates, while cells from the marginal regions form endodermal and mesodermal derivatives. A more complex pattern is observed for the fate map along the presumptive dorsoventral axis; dorsal animal cells will form neuroectoderm, while ventral animal cells will form epidermal structures like skin and fins. Of the presumptive mesoderm in the marginal zone, dorsal-most cells will form the hatching glands and the notochord, dorsolateral cells the heart and somites, ventrolateral cells the pronephros, and ventral-most cells will form blood. In this chapter, we will study how positional values within the early zebrafish embryos are established and interpreted by the respective cells, which genes are involved in these processes, and how the various gene products interact.

6.2 What We Know from Studies in the Frog

Early dorsoventral patterning processes have been intensively studied in amphibian embryos (Fig. 1). Under maternal control, which means governed by factors that have been generated by the mother and deposited in the egg, a coarse initial dorsoventral pattern is set up in the developing mesoderm. The mesoderm is induced in the equatorial/marginal region of the amphibian blastula by signals emanating from vegetal blastomeres which themselves will form endoderm. Ventral mesoderm, which gives rise to blood, is induced in a broad equatorial region spanning most of the dorsoventral axis, while dorsal mesoderm, which gives rise to the notochord, is induced in a rather small dorsal equatorial domain. This initial dorsoventral pattern is subsequently refined under the control of zygotic signals generated by the embryo itself. Dorsalizing signals from the dorsal mesoderm, also called the Spemann organizer, have two functions: firstly, they convert mesoderm in the dorsolateral equatorial region, initially specified as ventral, into a whole spectrum of intermediate fates, so that instead of blood, somitic muscle or pronephros is formed; secondly, they induce neural specification in dorsal animal cells which would otherwise give rise to epidermal fates. These dorsalizing signals are counteracted by zygotic ventralizing signals generated in the rest of the embryo. Here, we will deal with the second phase of dorsoventral patterning, the zygotic refinement of the initial, coarse dorsoventral pattern.

Several candidates for the dorsalizing and ventralizing signals have been identified by different means: the bone morphogenetic proteins Bmp2, Bmp4, and/or Bmp7 – members of the TGFβ superfamily of growth factors, which have ventralizing activities and , when applied to amphibian or fish embryos, lead to an expansion of ventral fates at the expense of dorsal fates – and their antagonists Chordin, Noggin, and Follistatin (for reviews see Harland and Gerhart 1997; Heasman 1997; Thomsen 1997) – which have dorsalizing activities and , when applied to embryos, lead to an expansion of dorsal fates at the expense of ventral

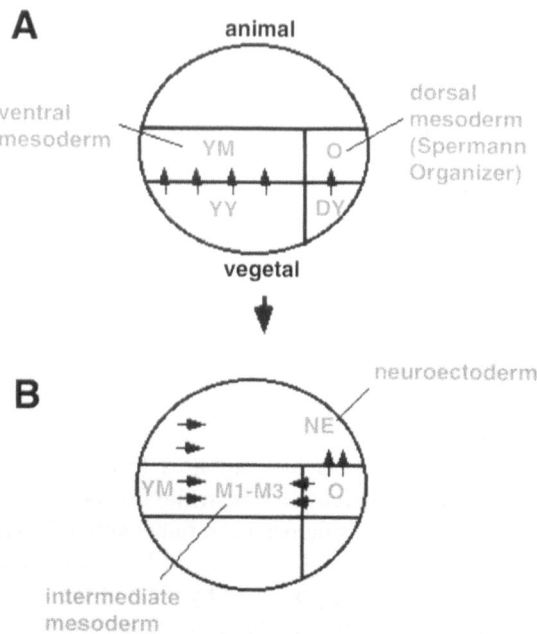

Fig. 1A,B. Four-signal model summarizing inductive processes during dorsoventral pattern formation of the amphibian embryo.**A** Maternally supplied signals emanating from vegetal cells induce mesoderm in the marginal zone of the early blastula embryo. Signals from ventral vegetal cells (*VV*) induce ventral mesoderm (*VM*) in a broad marginal region, while signals from dorsal vegetal cells (*DV*) induce dorsal mesoderm, also called the Spemann organizer (*O*), in a comparably small dorsal marginal domain. **B** The initial dorsoventral pattern is refined under the control of zygotic signals generated by the embryo itself. Dorsalizing signals from the Spemann organizer (*O*) convert the initially ventrally specified mesoderm of dorsolateral marginal positions into more intermediate fates (*M1-M3*). In addition, they induce neural specification in animal dorsal regions of the embryo which would otherwise give rise to epidermal fates. Both inductions of Spemann organizer signals are antagonized by ventralizing signals generated in the rest of the embryo. *VM* will form blood, *M3* muscle of the anterior somites, and *O* notochord

fates. Recent evidence indicates that Bmp4 can function as an instructive morphogen which determines positional identities along the entire dorsoventral axis in a dose-dependent fashion (Dosch et al. 1997), while Chordin, Noggin, and Follistatin attenuate this ventralizing activity on the dorsal side of the embryo by physical interaction with, and inhibition of, Bmp proteins (Fainsod et al. 1997; Piccolo et al. 1996; Zimmerman and Harland 1996), thereby leading to the establishment of the putative dorsoventral gradient of Bmp4 activity. According to this model, the specification of dorsal-most fates occurs in a kind of default pathway, when Bmp activity is entirely abolished, while intermediate Bmp levels lead to the establishment of intermediate fates like muscle, and highest Bmp levels to the establishment of ventral-most fates such as blood.

Final evidence, however, that early dorsoventral patterning does indeed occur according to this model, as well as the identification of the involved genes, requires genetical analyses via mutant embryos in which the activity of the respective genes is reduced (hypomorphic mutations) or entirely lost (amorphic mutations).

6.3 The Genetics of the Fish

Hypomorphic and amorphic mutants have been isolated in the zebrafish. Here, large-scale screens for mutants with defects in various developmental processes have been carried out. After random introduction of point mutations via the chemical ethylnitrosourea (ENU) and classic inbreeding steps, homozygous mutant embryos were generated (Haffter et al. 1996; Mullins et al. 1994). Thereby, thousands of zygotic recessive mutations were isolated and assigned to several hundred complementation groups; among them at least six complementation groups defining six genes zygotically required for ventral development and two complementation groups defining two genes zygotically required for dorsal development (Hammerschmidt et al. 1996a; Mullins et al. 1996; Solnica-Krezel et al. 1996). At this point of their identification, these genes are defined only by the phenotype they cause in the embryo upon mutation. Their molecular nature, however, is not known a priori, and has to be determined in further steps.

6.4 Morphology of Dorsalized and Ventralized Zebrafish Mutants

The phenotypes of zebrafish mutants, with specific defects in dorsal or ventral specification during early dorsoventral pattern formation, could already be predicted from the effects caused by the overexpression or specific inhibition of the ventralizing agents Bmp2, Bmp4, or Bmp7 in genetically wild-type embryos. Overexpression of Bmps was achieved by microinjecting different amounts of synthetic *bmp* mRNAs into zebrafish embryos of the 1-cell stage, inhibition of Bmps by microinjection of mRNAs encoding one of their aforementioned inhibitors, Noggin, Chordin, or Follistatin (Bauer et al. 1998; Kishimoto et al. 1997). At 36 h postfertilization, embryos injected with the lowest amount of *noggin* mRNA display a partial loss of the ventral tail fin, the ventral-most ectodermal derivative, defining the mildest degree of dorsalization, classified as C1. Increasing amounts of injected *noggin* mRNA lead to a complete loss of the ventral tail fin, accompanied by a winding up of the body axis which is restricted to the tip of the tail in cases of mild dorsalization (C2), but affects the entire tail in cases of a dorsalization of intermediate strength (C3), and both tail and trunk in cases of strongest dorsalization (C4-C5). From C3 dorsalization upwards, a loss of the blood islands, the ventral-most mesodermal derivate, is observed. The winding up of the body axis is a consequence of a lateral expansion of the somites which occurs only in the posterior-most in cases of mild dorsalization, but progressively affects more anterior somites in more strongly dorsalized embryos. In cases of strongest dorsalization (C5), even the anterior-most somites have spread into ventral-most regions and are fused on the ventral side of the embryo, which leads to a rupture of the yolk sac upon somite constriction around the 15-somite stage.

A similar continuous series of phenotypes of increasing strengths is observed upon the injection of increasing amounts of Bmp mRNAs. As in the case of Bmp inhibition, the most sensitive structure responding to lowest amounts of injected *bmp* mRNA is the ventral tail fin, which displays a partial or complete duplication (V1). Increasing amounts of *bmp* mRNA lead to a progressive enlargement of the blood islands and the tail, and a progressive reduction in the size of head and eyes, resulting from a progressively impaired neural induction (V2,V3). Most strongly ventralized embryos are characterized by a complete absence

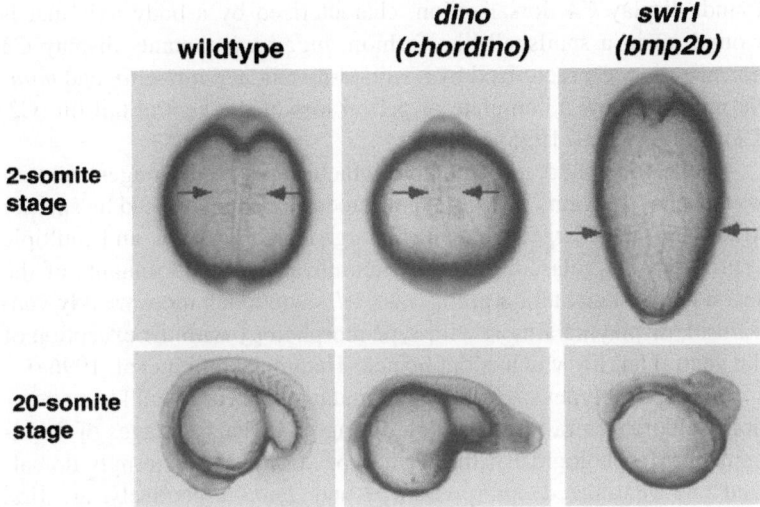

Fig. 2. Morphology of *chordino* and *swirl/bmp2b* mutant embryos at 2-somite (*upper row*, dorsal view anterior up) and 20-somite stage (*lower row*, lateral view anterior to the left). *Arrows* in the *upper row* point to the first somites as a reference point indicating the shifts in anteroposterior organization. At the 2-somite stage, *chordino* mutants display an enlargement of the posterior region at the expense of the anterior region. In addition, the anterior somites are smaller and the posterior region of the notochord is missing, both derivatives of the dorsolateral mesoderm. *swirl/bmp2b* mutants, in contrast, display an enlargement of the anterior region at the expense of the posterior region. In addition, the notochord anlage is broadened, and the anterior somites are laterally expanded and fused on the ventral side of the embryo. At the 20-somite stage *chordino* mutants display an enlargement of the tail with enlarged blood islands, while the size of the head is reduced. In *swirl/bmp2b* mutants the body axis is wound up in a snailshell-like fashion

of all head structures and a partial loss of the somites and notochord, both derivatives of the dorsolateral mesoderm (V4).

During our mutant screen, we found several mutants resembling those generated by Bmp overexpression or inhibition, whose phenotypic strengths could be classified accordingly. Mutants of the *swirl* complementation group, of which two alleles were isolated, display the strongest dorsalization (C5) and lyse around the 15-somite stage; *somitabun* and *snailhouse* mutants, for each of which a single allele was found, display C4 dorsalization, characterized by a body axis that is wound up in a snailshell-like fashion. *piggy tail* mutants display C3 dorsalization, characterized by a wound-up tail, and *lost-a-fin* and *mini-fin* mutants show a complete or partial loss of the ventral tail fin (C2, C1) (Mullins et al. 1996).

On the other hand, *dino* mutants, which display the strongest ventral-ized phenotype found in the mutant screen, are characterized by smaller heads and an enlarged tail with enlarged blood islands and multiple ventral tail fins, classified as V3 ventralization, while mutants of the second complementation group, *mercedes*, are much more weakly ven-tralized, displaying almost wild-type morphology with the exception of the ventral tail fin which is duplicated (Hammerschmidt et al. 1996a).

These phenotypes of 36-hour-old mutant embryos result from altera-tions in dorsoventral organization during much earlier stages of devel-opment. Morphologically, the phenotypes of the most strongly dorsal-ized and ventralized embryos, *swirl* and *dino*, respectively, are first visible at midgastrula stages. At the 2-somite stage (Fig. 2; see also Hammerschmidt et al. 1996b), *dino* mutants display a reduction in the size of anterior somites and posterior notochord, both derivatives of the dorsolateral mesoderm, while both structures are greatly enlarged in *swirl* mutant embryos. In addition to these dorsolateral alterations, both mutants display shifts in the anteroposterior organization. In *dino* mu-tants, the posterior region is enlarged at the expense of the anterior region, in *swirl* mutants, the anterior region is enlarged at the expense of the posterior region (Fig. 2). This tight linkage of dorsal and anterior fates on one hand, and ventral and posterior fates on the other hand, results from the morphogenetic movements during gastrulation, epi-boly, dorsal convergent extension, and involution, which let dorsal cells end up in more anterior positions of the developing body axis than cells that come from ventral regions of the pregastrula embryo.

6.5 Analysis of Dorsoventral Phenotypes with Molecular Markers

A detailed analysis of *dino* and *swirl* mutant embryos was carried out using molecular markers that help to identify cell fates long before the specification of cells becomes morphologically visible. At late blastula stages, the expression pattern of *bmp2b* and *chordino* were unaffected in both mutants, indicating that the initial coarse dorsoventral pattern, which is supposed to be under maternal control (see above), is set up normally, including a normal induction of the Spemann organizer. However, severe alterations are observed beginning shortly before the onset of gastrulation, when the initial pattern is supposed to be refined under the control of zygotic signals from the Spemann organizer and antagonizing ventralizing signals. In normal embryos, *eve1*, a marker for presumptive ventral mesoderm (Joly et al. 1993), is initially expressed in a broad fashion. Beginning shortly before the onset of gastrulation, its expression becomes progressively restricted to ventral-most regions, indicating the progressive dorsalization of mesoderm in dorsolateral regions initially specified as ventral. In *dino* mutants, this ventral retraction of the *eve1* expression domain does not occur, whereas in *swirl* mutants, *eve1* expression is progressively lost in the entire embryo, including ventral-most regions. On the other hand, the expression domain of *fkd3*, a marker for presumptive neuroectodermal cells in dorsal animal regions (Hammerschmidt et al. 1996a), is much narrower in *dino* mutants, but expanded into ventral-most regions in *swirl* mutants. In an animal view, the expression domains of the ventral mesodermal marker *eve1* and the neuroectodermal marker *fkd3* are perfectly complementary in wild-type, *dino,* and *swirl* mutants, indicating that dorsalization of the mesoderm and neural induction are indeed tightly linked, and most likely regulated by the same signals that are affected in *dino* and *swirl* mutants. *dino* mutants appear to display a defect in a dorsalizing signal generated from the Spemann organizer, while *swirl* mutants apparently have a defect in an antagonizing signal required for the maintenance of ventral specification. The phenotype of *snailhouse* and *somitabun* mutants is very similar to that of *swirl* mutants, suggesting that the same processes are affected.

6.6 Early Expression Patterns of *bmp2b*, *bmp4*, *bmp7*, *chordino,* and *follistatin*

We were interested to learn which genes might be affected in the various dorsoventral patterning mutants. Before starting laborious approaches such as positional cloning (see for example, Zhang et al. 1998), we

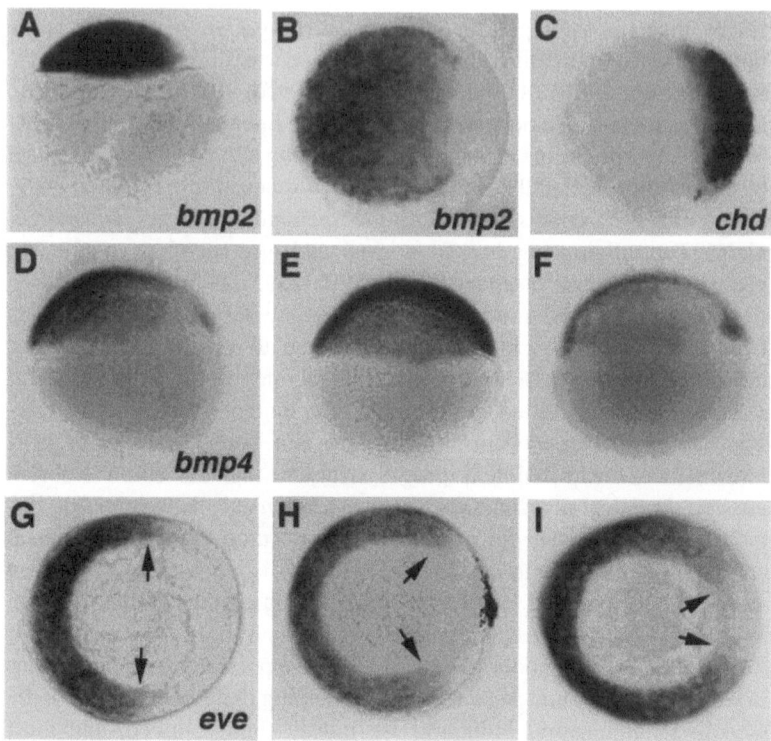

Fig. 3A-I. (**A-F**) Expression patterns of *bmp2b*, *chordino (chd),* and *bmp4*, detected via whole mount in situ hybridization. Dorsal is always to the right, ventral to the left. **A,B** Initial expression patterns of *bmp2b* (**A** lateral view, **B** animal view) and (**C**) *chordino* (animal view) at a late blastula stage, 2 h before the onset of gastrulation. *bmp2b* transcripts are uniformly distributed in the embryo with the exception of the dorsal marginal region, the presumptive Spemann organizer, which shows expression of *chordino*. Expression pattern of *bmp4* at the onset of gastrulation, lateral view, in a wild-type (**D**), a *chordino* mutant (**E**), and a *swirl/bmp2b* mutant (**F**) embryo. Compared to wild-type (**D**), the *swirl/bmp2b* mutant (**F**) displays a significant reduction in *bmp4* transcription, indicating that *bmp4* is a target of Bmp2 signaling, and that *bmp4* mRNA levels reflect Bmp2b activity. In the wild-type embryo (**D**), *bmp4* displays a graded expression pattern with mRNA levels progressively dropping from ventral to dorsal, reflecting a gradient of Bmp2b activity. In the *chordino* mutant (**E**), *bmp4* displays a uniform expression throughout the embryo, indicating that instead of the establishment of a gradient of Bmp2b activity, the initial broad and uniform pattern is maintained. **G-I** Long-range signaling of Chordino, revealed in a chimeric embryo after transplantation of a few (*darkly labeled*) wild-type cells into the Spemann organizer of a *chordino* mutant embryo (**H**). Whole mount in situ hybridization of wild-type control (**G**), chimeric (**H**), and *chordino* mutant control embryo (**I**) with *eve1* probe (Joly et al. 1993), marking presumptive ventral mesodermal cells at a midgastrula stage, 2 h after the transplantation. View from vegetal pole, dorsal to the right. The dorsal boundary of the *eve1* expression domain is marked by *arrows*. In the wild-type embryo (**G**), *eve1* expression is restricted to the ventral side, while in the *chordino* mutant, the *eve1* expression domain is extended far into the dorsal side. Transplantation of a few wild-type cells leads to a significant ventral retraction of the extended *eve1* expression domain. Lateral regions many cell diameters away from the wild-type cells, indicate that Chordino signaling has a long-range effect, spreading over approximately a quarter of the entire dorsolateral axis within 2 h

tested several candidate genes previously isolated from other organisms where they show activities like those proposed for the gene products affected in the zebrafish mutants. As ventralizing signals which might be affected in dorsalized mutants, three candidates had been isolated, the Bone Morphogenetic Proteins Bmp2, Bmp4, and Bmp7, as dorsalizing signals which might be affected in the ventralized mutants, the Bmp antagonists Noggin, Chordin, and Follistatin. The zebrafish homologues of these genes were cloned (Bauer et al. 1998; Martínez-Barberá et al. 1997; Miller-Bertoglio et al. 1997; Nikaido et al. 1997; Schulte-Merker et al. 1997) and their expression patterns investigated. Two of the *bmp* genes, *bmp2b* and *bmp7*, are expressed from blastula stages onwards in a broad ventral domain, reflecting the initial broad ventral specification within the zebrafish blastula (see Fig. 3A,B for *bmp2b*). The expression of *bmp4* starts later, shortly before the onset of gastrulation. During early gastrula stages, *bmp2b*, *bmp4*, and *bmp7* show a very similar, graded expression pattern with mRNA levels progressively dropping from ventral to dorsal (see Fig. 3D for *bmp4*). During further development, however, the expression patterns of the three genes diverge. At late gastrula stages, *bmp2b* shows the broadest ventral expression domain, sparing just the embryonic axis, while *bmp7* is expressed in the ventral half of the embryo, and *bmp4* only in a rather small ventral posterior domain. In addition to this ventral expression, all three *bmp* genes are differentially expressed in the dorsal midline of the embryo, the prechordal plate, or the notochord anlage. During somitogenesis stages, *bmp2b* is initiated in the eyes and the otic vesicles, *bmp4* in the eyes, the head region, the dorsal ectoderm, the lateral mesoderm, and in ventral regions of the somites. *bmp7* is expressed in a rather small region of the brain, probably the pineal gland, in addition to strong expression remaining in ventral regions of the embryo. Of the dorsalizing candidates, only the zebrafish *chordin* homologue, named *chordino*, is expressed at the right time and place to be a good candidate for the Spemann organizer signal affected in *dino* mutants. Like *bmp2* and *bmp7*, *chordino* expression starts at blastula stages, when it is expressed complementarily to *bmp2b* and *bmp7* in the region of the presumptive Spemann organizer (Fig. 3C). The expression of *follistatin* starts later, shortly after the onset of gastrulation when it is expressed in the presumptive anterior paraxial mesoderm. The expression pattern suggests that Follistatin might act as an antagonist of Bmp2b during later stages of dorsoventral patterning. However, it can be ruled out as a Spemann organizer signal of the zebrafish embryo.

6.7 *dino, swirl,* and *snailhouse* Are *chordino, bmp2b,* and *bmp7*

As a first step of the candidate testing approach, we mapped the loci of candidate genes and of the dorsoventral mutations, or carried out a direct linkage analysis to test whether genes and mutations cosegregate during meiosis. For this purpose, a polymorphism in the candidate gene has to be identified which allows the gene of the mutant line to be distinguished from that of a wild-type reference line the mutant is crossed into. In the case of *chordino* and *dino*, linkage analysis was carried out using an EcoR1 restriction fragment length polymorphism (RFLP) present between the *dino* mutant line and the wik-reference line (Schulte-Merker et al. 1997). F1 hybrids from a cross of a *dino* carrier and a wik fish were raised, *dino* carriers identified and set up with each other. The resulting F2 embryos were sorted into wild-type and mutant pools which were analyzed for the *chordino* RFLP in genomic Southern blots. Analysis of over 200 mutant embryos revealed complete absence of the wik-specific *chordino* allele, indicating that the *chordino* gene and the *dino* mutation are linked and located within 0.25 cM from each other (no recombination in 400 meioses). Cloning of the *chordino* cDNA from one of the *dino* alleles, din^{tt250}, revealed a 104 bp deletion at the 5' end of the coding region, which only leaves the first 40 amino acids of the original protein. This deletion causes a frameshift which leads to a premature termination of the protein after an additional 50 amino acids. Since these alterations occur upstream of the first functional, cysteine-rich domain, din^{tt250} Chordino is most likely a functional null mutant.

Similar analyses, in this case using a single strand conformational polymorphism (SSCP) approach, revealed that the *swirl* mutation is linked to the zebrafish *bmp2b* gene (Kishimoto et al. 1997). Here, two alleles of similar strength were sequenced. In one of them, swr^{tc300}, a conserved cysteine residue normally involved in the formation of intramolecular disulfide bonds, is replaced by a tryptophane, while in the other, swr^{ta72}, Bmp2b protein is C-terminally extended by six additional amino acids.

Finally, *snailhouse* was shown to map within 0.16 cM of zebrafish *bmp7* (no recombination in over 300 mutant F2 embryos tested=600 meioses using a SSCP). Sequencing of the only available allele, snh^{ty68}, revealed a glycine-valine exchange in the signal peptide of Bmp7 protein which appears to abolish the secretion of the mutant protein. Upon overexpression in zebrafish embryos and *Xenopus* animal cap explants, this mutant form of Bmp7 revealed a severely reduced ventralizing activity. The same was true for *Xenopus* Bmp7 after introduction of the same mutations via site-directed mutagenesis. *bmp4, follistatin* and *noggin* did not map to any of the other dorsoventral mutations.

Altogether, these results indicate that *bmp2b* and *bmp7* are essential for ventral development, and that *chordino* is essential for dorsal development during early dorsoventral patterning of the zebrafish embryo. However, in contrast to what is suggested by their activity in *Xenopus* animal cap assays, both *bmp2b* and *bmp7* are not required for the initial induction of mesoderm. Interestingly, targeting of the mouse *bmp2, bmp7,* and *chordino* homologues did not affect early dorsoventral patterning but later processes of mouse development. *bmp2*-deficient mice, for example, display defects in amnion and heart development (Zhang and Bradley 1996), while *bmp7*-deficient mice show defects in eye and kidney formation (Dudley et al. 1995; Luo et al. 1995). On the other hand, these later processes do not appear to be affected in the zebrafish mutants. The dorsoventral defects of *bmp2b, bmp7,* and *chordino* mutants can be rescued via injection of the respective mRNAs at the 1-somite stages. Rescued individuals are viable and can be raised to fertile adults, although injected RNA is usually degraded by the end of gastrulation (Hammerschmidt et al. 1998). The analysis of rescued homozygous mutant adults revealed no apparent alterations in the case of *snailhouse/bmp7* and *chordino*, while *swirl/bmp2b* mutants displayed rather severe balancing problems. Two organs have been described to be involved in balancing and sensing the orientation in the fish, the lateral line and the inner ears. *bmp2b* consistently shows a strong expression in the otic vesicles during late stages of somitogenesis. A more thorough analysis of the inner ear defects in rescued *swr bmp2b* mutant adult fish is currently under way.

6.8 Chordino Acts via Bmp2b and Bmp7

To test whether Chordino functions via an inhibition of ventralizing proteins of the Bmp family, as suggested by biochemical analyses carried out in amphibia (Piccolo et al. 1996), *chordino-bmp2b* (Hammerschmidt et al. 1996b) and *chordino-bmp7* double mutants were generated. Such double mutants appear dorsalized much like *bmp2b* and *bmp7* single mutants, indicating that Bmp2b and Bmp7 are epistatic to Chordino. This suggests that Chordino is not an actively dorsalizing agent, but does indeed act as an inhibitor of Bmp2b and Bmp7. Clearly, the absence of such an inhibitor has no further effects when its substrate is missing anyway.

The analysis of the *bmp2b*, *bmp4*, and *bmp7* expression patterns in *chordino*, *bmp2b*, and *bmp7* mutant embryos suggests that Chordino is required for the establishment of the putative morphogenetic Bmp2/4/7 gradient. *bmp2* and *bmp7* are initially expressed in a uniform and broad fashion, sparing only the region of the presumptive Spemann organizer where *chordino* is expressed. During further development, this pattern is transformed into a graded pattern (see above and Fig. 3). In *chordino* mutants, however, the initial broad and uniform expression pattern is maintained (see Fig. 3E for *bmp4*), indicating that Chordino is required for the progressive clearing of *bmp2/4/7* expression from dorsal to ventral. This clearing is achieved indirectly via the inhibition of Bmp2/4/7 activity. Bmps are positive autoregulators of their own expression, as indicated by the progressive loss of *bmp2/4/7* transcripts in *swr bmp2b* and *snh bmp7* double mutant embryos (see Fig. 3F for *bmp4* in *swr*). Thus, by binding and inhibiting Bmp proteins, Chordino attenuates this positive autoregulation, leading to decreased *bmp2/4/7* transcription. Because of the positive autoregulation, the levels of *bmp2/4/7* transcripts can be regarded as a readout of Bmp2/4/7 activity. Therefore, the observed gradient of *bmp2/4/7* transcripts could be a direct reflection of the proposed gradient of Bmp2/4/7 activity.

6.9 Long-Range Effects of Chordino Signaling

The rearrangement of the *bmp2b* and *bmp7* mRNA distribution from the uniform to the graded pattern takes place in the entire embryo, including ventral-most regions very distant from the Spemann organizer, the site of Chordino production. Thus, if Chordino function was indeed responsible for this transformation, one has to propose that Chordino has long-range signaling effects. To test this hypothesis, chimeric embryos were generated by transplanting labeled wild-type cells into *chordino* mutant host embryos (Hammerschmidt et al. 1996b). When wild-type cells were transplanted into the future ventral side of a late blastula *chordino* mutant embryo, the mutant phenotype was maintained at normal strength, whereas transplantation into the future dorsal side led to a striking rescue of the phenotype to almost wild-type condition. This effect was achieved even when only a few cells (approximately 20) were transplanted. Although after 48 h of development, the descendants of these cells were all located in very anterior positions, the chimeric embryo displayed a rescue of the *chordino* mutant phenotype over the entire length of its body axis, indicating that Chordino must act in a non-cell autonomous fashion with long-range effects, influencing even the fate of very distant cells. With molecular markers, this long-range effect of the Chordino-generating wild-type cells can already be revealed at midgastrula stages, a few hours after cells had been transplanted (Fig. 3G-I). Wild-type cells were transplanted directly into the Spemann organizer of an early gastrula mutant embryo. After 2 h, the wild-type cells were located in the presumptive notochord anlage. Such chimeric embryos displayed a significant retraction of the expression of the ventrolateral marker gene *eve1* into lateral regions. Even mutant cells many cell diameters away from the wild-type cells, the source of functional Chordino protein, had lost their *eve1* mRNA expression (Fig. 3H). No such effect was observed upon transplantation of *chordino* mutant cells. This indicates that Chordino signaling can spread over about a quarter of the entire dorsoventral axis of an early gastrula embryo within 2 h.

It is currently unclear whether this long-range effect is caused by Chordino proteins directly, or whether secondary signals are involved. In any case, the long-range signaling, which occurs on the level of the protein, is further potentiated by a progressive lateral expansion of the *chordino* expression domain. Unlike in *Xenopus*, expression of *chordino* does not remain restricted to the dorsal mesoderm, but starts to spread into more lateral regions shortly after the onset of gastrulation. This spreading of the *chordino* expression appears to occur in a kind of progressive homogenetic induction, as it is not seen in *chordino* mutants, where *chordino* expression remains restricted to the dorsal-most region of the embryo (Schulte-Merker et al. 1997). This indicates that the lateral spreading requires functional Chordino protein, and that during normal development, Chordino acts as an inducer of its own expression in neighboring cells. This positive autoregulation is achieved indirectly via an inhibition of Bmp2/4/7, as suggested by the effects of *bmp2/4/7* overexpression and BMP inhibition on induced *chordino* expression.

In summary, the ventral retraction of *bmp2/4/7* expression and the lateral spreading of the expression of *chordino* appears to occur by the following mechanism: Bmp2/4/7 act as positive regulators of their own expression, as described above, and as negative regulators of *chordino* expression. Chordino protein, when diffusing from its site of production, binds and inactivates BMP proteins, which has two synergistic effects on the transcriptional level in the neighboring cells: a repression of *bmp2/4/7* transcription and a de-repression of the transcription of *chordino*.

6.10 Tolloid/Bmp1 Acts as an Inhibitor of Chordino

Having learned that Chordino is a rather efficient signaling molecule and positive autoinducer, it is tempting to imagine that the embryo has developed means to restrict Chordino action and to protect ventral-most regions from Chordino signaling. A candidate for such a potential inhibitor of Chordino named Tolloid has been recently isolated in the fruit fly *Drosophila* (Finelli et al. 1994), *Xenopus* (Piccolo et al. 1997), and zebrafish (Blader et al. 1997). A very similar protein had been previously isolated in mammals, named Bone Morphogenetic protein Bmp1. Unlike the other Bmp proteins, Bmp1 is not a TGFβ molecule, but a metalloprotease of the astacin family. It was designated as a Bone Morphogenetic protein, because it was originally identified together with Bmp2 and Bmp3 from demineralized bone that induced ectopic

cartilage and endochondral bone when implanted in experimental animals (Wozney et al. 1988). Bmp1 is supposed to cleave procollagen and
to facilitate collagen assembly into fibers during bone formation
(Kessler et al. 1996), although *bmp1*-deficient mice do not display
major skeletal abnormalities (Hogan 1996).

During early zebrafish development, *tolloid* is expressed on the ventral side of the midgastrula embryo. After gastrulation, strong *tolloid*
expression is maintained in the forming tailbud, very similar to the
expression of *bmp4* (Blader et al. 1997). In vitro, wild-type Tolloid
protein leads to a specific degradation of bacculovirus-generated Chordin protein (Blader et al. 1997; Piccolo et al. 1997). As in the case of
bmp2b and *bmp7, tolloid* could be assigned to a dorsalized zebrafish
mutant. Work carried out in the laboratory of Mary Mullins at the
University of Pennsylvania, Philadelphia, indicates that the mild C1
dorsalization of *minifin* mutants is caused by potential null mutations in
the zebrafish *tolloid* gene (Conners et al. 1999). *Tolloid-chordino* double mutants are ventralized like *chordino* single mutants, indicating that
chordino is epistatic to *tolloid* and that Tolloid does indeed function as
an inhibitor of Chordino. The relatively mild dorsalization of *tolloid*
mutants, however, suggests that Chordino inhibition by Tolloid is not of
central importance for dorsoventral patterning of the zebrafish embryo,
but only required in ventral-most regions, unless there are partially
redundant genes that inhibit Chordino in concert with Tolloid.

6.11 Zebrafish *smad1* and *smad5*

So far, we have dealt with the mechanisms that define positional values
along the dorsoventral axis of the pregastrula embryo by establishing
the putative Bmp2/4/7 gradient. However, we would also like to know
how this gradient and the local Bmp2/4/7 concentrations are interpreted
by the different cells along the dorsoventral axis. Bmp2 and 4 are known
to bind to transmembrane receptors of the serine-threonine kinase receptor family which, upon ligand binding, phosphorylate and activate
Smad proteins. For the mediation of BMP signaling, three different
Smad proteins have been identified: Smad1, Smad5, and more recently,
Smad8. These receptor-regulated Smad proteins usually form cytoplasmic homotrimers which, after their receptor-mediated phosphorylation,

associate with Smad4 trimers and translocate to the nucleus where they participate in transcriptional complexes (see Attisano and Wrana 1998, Kretschmar and Massagué 1998, for reviews). We have cloned the zebrafish homologues of Smad1 and Smad5.

Developmental Northern analysis reveals that high levels of *smad5* transcripts are provided maternally. At early blastula stages, *smad5* mRNA levels have dropped to about 50%. Levels go up again after the onset of zygotic transcription at midblastula stages, but decline soon afterwards. *smad1*, on the other hand, starts to be expressed at gastrula stages, long after the peak of *smad5* expression. Whole mount in situ hybridization revealed that both maternal and zygotic *smad5* transcripts are uniformly and ubiquitously distributed, whereas the expression of *smad1* is restricted to the ventral side of the gastrula embryo. This expression is absent in *bmp2b* mutant embryos, but enhanced after overexpression of *bmp2b* in wild-type embryos, indicating that *smad1* transcription is positively regulated by Bmp2b signaling.

6.12 *somitabun* Is *smad5*

Zebrafish *smad1* and *smad5* genes were mapped, revealing that *smad5* maps close to another dorsalizing mutation, *somitabun sbn^{tc24}*. Subsequent direct linkage analysis between the *smad5* gene and the *somitabun* mutation revealed no recombination in over 200 examined meioses, indicating that gene and mutation are located within 0.5 cM of each other. Consistent with the maternal and zygotic expression of *smad5*, the *somitabun* mutation has both a dominant maternal and a dominant zygotic effect. When a heterozygous somitabun female is crossed to a wild-type male, all offspring show strong dorsalization of C4 strength, characterized by a wound-up trunk and tail. The observation that all offspring show the same phenotype, independent of their own genotype (either wild type or heterozygous for *sbn*) indicates the overwhelming maternal effect of the mutation. However, *sbn* also has a weak zygotic effect. When heterozygous males are crossed to wild-type females, 50% of the offspring show a weak C1 dorsalization, characterized by the absence of the ventral tail fin.

Cloning of the *smad5* cDNA from *somitabun* mutant embryos revealed a single amino acid exchange, a threonine to isoleucine exchange in the L3 loop region of the MH2 domain of Smad5 protein. This threonine residue is conserved in all currently known receptor-regulated Smad protein. Mutational in vitro analyses carried out for mouse Smad2 have shown that this threonine residue is involved in the binding of receptor-regulated Smad proteins to the cytoplasmic domain of the Bmp/TGFβ receptors and to Smad4, while the formation of ho- motrimers is not affected. This might explain the dominant negative, antimorphic activity of the *sbn* Smad5 mutant, assuming that mutant protein traps wild-type Smad5 in inactive trimers which cannot bind to the Bmp/TGFβ receptors or to Smad4.

When mutant *smad5* mRNA is injected into wild-type embryos, the *sbn* mutant phenotype (C4 dorsalization) is phenocopied, while injec- tion of high amounts of wild-type *smad5* mRNA into *sbn* mutant em- bryos leads to a rescue of the mutant phenotype to wild-type condition. In addition to mRNA, phenocopy and rescue experiments were also carried out with plasmid DNAs that drive *smad5* expression under the control of two promoters with different temporal activation profiles; the *Xenopus* EF1α promoter, which is activated right after midblastula transition when zygotic gene expression starts (Kane and Kimmel 1993), and the cytoskeletal actin (CSKA) promoter, which is strongly activated during gastrulation, but very weakly expressed at earlier stages (Hammerschmidt, et al. 1998). Injection of wild-type *smad5* DNA into *sbn* embryos only led to a rescue, and injection of mutant *smad5* DNA into wild-type embryos to a phenocopy of the *sbn* pheno- type, when gene expression was under the control of the EF1α pro- moter; no effect was observed upon injection of the corresponding CSKA constructs. These data suggest that *smad5* acts after midblastula transition and before gastrulation.

The antimorphic nature of the *sbn smad5* mutation was also con- firmed in *Xenopus* animal cap explants. Wild-type *smad5*, when overex- pressed in animal caps, induces the expression of ventral mesodermal markers like *Xhox3*. Upon injection of *sbn smad5*, however, no induc- tion of *Xhox3* transcription is observed, indicating that *sbn* Smad5 has retained no, or very weak, ventralizing activity. Coexpression of wild- type and mutant *smad5* revealed that the ventralizing effect of wild-type *smad5* RNA was inhibited approximately 80% by equal amounts of co-injected *sbn*[tc24] *smad5* RNA, and 90% by twofold amounts.

6.13 Smad5 Is Involved in the Mediation of Early Bmp2b Signaling

To assess the relationship between *smad5* and *bmp2b*, early marker gene expression patterns were compared between *sbn smad5* and *swr bmp2b* double mutant embryos. Despite the maternal effect of the *sbn* mutation and the maternal expression of *smad5*, no alterations in dorsoventral patterning of *sbn* mutant embryos were observed earlier than in *bmp2b* mutant embryos, indicating that Smad5 might have no function prior to the function of the zygotically generated signal Bmp2b.

The phenotype of *sbn smad5–swr bmp2b* double-heterozygous embryos indicates that Bmp2b and Smad5 interact. As described above, the zygotic effect of the *sbn smad5* mutation leads to a weak C1 phenotype. *sbn-swr* double-heterozygous embryos deriving from a cross of a *sbn* heterozygous male and a *swr* heterozygous female, however, are much more strongly dorsalized and display C3 dorsalization, although heterozygosity for *swr* alone has no effect at all. A similar enhancement is also observed for the maternal effect of the *sbn smad5* mutation (from C4 to C5).

Epistasis analyses via double mutants to determine whether Smad5, as expected of a signal transducer, does act downstream of Bmp2b are not possible, since mutations in both genes generate the same type of phenotype. Instead, the effect of overexpressed Bmp2/4 in *sbn smad5* mutants was studied. At the onset of gastrulation, *sbn smad5* mutant embryos display a striking reduction in *bmp2b* mRNA levels, similar to the phenotype of *swr bmp2b* mutants themselves. Injection of wild-type *smad5* mRNA into *sbn* mutant embryos leads to a rescue of *bmp2b* mRNA levels to wild-type condition, whereas the injection of *Xenopus bmp2* or *bmp4* mRNA showed no effect. This indicates that Smad5 does indeed act downstream of Bmp signaling to mediate the aforementioned positive autoregulation of *bmp2/4* expression.

6.14 The Three Phases of Dorsoventral Patterning

In contrast to their unresponsiveness in early *bmp2b* expression, *bmp2/4*-injected *sbn* mutant embryos showed a striking response in their morphology during later stages of development. When moderate amounts of zebrafish or *Xenopus bmp2* or *bmp4* mRNA were injected, the dorsalization could be significantly normalized (from C4 to C1), and even converted to a ventralization upon injection of high amounts of *bmp2/4* mRNA (C4 to V3).

To determine at what time point during development exogenous Bmp2/4 can override the requirement for *smad5*, human *bmp4* was expressed in *sbn smad5* mutant embryos under the control of the cytoskeletal actin promoter. In contrast to the corresponding *smad5* transgene (see above), expression of the human *bmp4* gene under this promoter led to a striking rescue and even ventralization of *sbn* mutant embryos, suggesting that the *smad5*-independent response of the zebrafish embryo to exogenous Bmp2/4 occurs during gastrula stages.

In the *bmp2/4* RNA injection experiments described above, the actual in vivo concentrations of Bmp2/4 protein are unknown. To investigate whether *sbn* cells can respond to Bmp2/4 under physiological conditions, cell transplantation experiments were carried out. Labeled *sbn smad5* mutant cells were transplanted into wild-type embryos. In all cases, donor cells gave rise to blood and ventral tail fin, ventrally derived tissues that are completely absent in *sbn* mutant embryos. The frequencies of ventral tissue contribution were similar, independently of whether donor cells were wild-type or *sbn* mutant, and whether they were transplanted at late blastula or early gastrula stages. Together with the Bmp2/4 injection experiments, these data indicate that in contrast to the early mediation of Bmp2/4 signaling, the final specification of ventral cell fates can occur in the absence of functional Smad5.

Altogether, the behavior of the *sbn smad5* and *swr bmp2b* mutations define three distinct phases of dorsoventral patterning. In the first phase, an initial coarse dorsoventral pattern is set up at a time when the future dorsal mesoderm, the equivalent of the amphibian Spemann organizer, is induced in a small dorsal marginal domain characterized by the expression of *chordino* and opposed by the expression of *bmp2b* in the rest of the embryo. This pattern, which is most likely set up under maternal control, is independent of Bmp2b and Smad5, although in *Xenopus*, Bmp2/4 and Smad1/5 have been discussed as candidate maternal components involved in the induction of ventral mesoderm (Graff et al. 1996).

In the second phase, which is under zygotic control and dependent on Bmp2b and Smad5, the initial dorsoventral pattern is refined, leading to the transformation of the broad and uniform expression of *bmp2b* to a graded pattern, with *bmp2b* mRNA levels progressively dropping from the ventral to dorsal side. This transformation is governed by the antagonizing action of Bmp2b and its mediator Smad5 on one side and the Bmp2b antagonist Chordino on the other side. Judging from in situ hybridization patterns in wild-type and mutant embryos, the establishment of this Bmp2b gradient, which is thought to have morphogenetic character-defining positional values and cell fates along the dorsoventral axis, appears completed by the onset of gastrulation.

In a third phase, which takes place during gastrulation, the Bmp2/4 gradient is interpreted and cell specifications occur according to the local Bmp2/4 concentrations. This phase is independent of *sbn* smad5, and it is currently unclear which alternative transcription factor mediating Bmp2/4 signaling might confer this independence of Smad5.

References

Attisano L and Wrana JL (1998) Mads and Smads in TGFβ signalling. Curr Opin Cell Biol 10:188–194

Bauer H, Meier A, Hild M, Stachel S, Economides A, Hazelett D, Harland RM, Hammerschmidt M (1998) Follistatin and Noggin are excluded from the zebrafish organizer. Dev Biol 204:488–507

Blader P, Rastegar S, Fischer N, Strähle U (1997) Cleavage of the BMP-4 antagonist Chordin by zebrafish Tolloid. Science 278:1937–1940

Conners SA, Trout J, Ekker M, Mullins MC (1999) The role of *tolloid/minifin* in dorsoventral pattern formation of the zebrafish embryo. Development 126: 3119–3130

Dosch R, Gawantka V, Delius H, Blumenstock C, Niehrs C (*1997*) Bmp4 acts as a morphogen in dorsoventral patterning in *Xenopus*. Development 124:2325–2334

Dudley AT, Lyons KM, Robertson EJ (1995) A requirement for bone morphogenetic protein-7 during development of the mammalian kidney and eye. Genes Dev 9:2795–2807

Fainsod A, Deißler K, Yelin R, Marom K, Epstein M, Pillemer G, Steinbeisser H, Blum M (1997) The dorsalizing and neural inducing gene *follistatin* is an antagonist of *BMP-4*. Mech Dev 63:39–50

Finelli AL, Bossie CA, Xie T, Padgett RW (1994) Mutational analysis of the *Drosophila tolloid* gene, a human BMP-1 homolog. Development 120:861–870

Graff JM, Bansal A, Melton DA (1996) *Xenopus* Mad proteins transduce distinct subsets of signals for the TGFβ superfamily. Cell 85:479–487

Haffter P, Granato M, Brand M, Mullins MC, Hammerschmidt M, Kane DA, Odenthal J, van Eeden FJM, Jiang Y-J, Heisenberg C-P, Kelsh RN, Furutani-Seiki M, Warga RM, Vogelsang E, Beuchle D, Schach U, Fabian C, Nüsslein-Volhard C (1996) The identification of genes with unique and essential functions in the development of the zebrafish, *Danio rerio*. Development 123:1–36

Hammerschmidt M, Blader P, Strähle U (1998) Strategies to perturb zebrafish development. In: Detrich HW, Westerfield M, Zon LI (eds) Methods in Cell Biology 59. Academic Press, San Diego, pp 87–115

Hammerschmidt M, Pelegri F, Mullins MC, Kane DA, van Eeden FJM, Granato M, Brand M, Furutani-Seiki M, Haffter P, Heisenberg C-P, Jiang Y-J, Kelsh RN, Odenthal J, Warga RM, Nüsslein-Volhard C (1996a) *dino* and *mercedes*, two genes regulating dorsal development in the zebrafish embryo. Development 123:95–102

Hammerschmidt M, Serbedzija GN, McMahon AP (1996b) Genetic analysis of dorsoventral pattern formation in the zebrafish: Requirement of a BMP-like ventralizing activity and its dorsal repressor. Genes Dev 10:2452–2461

Harland RM, Gerhart JC (1997) Formation and function of Spemann's organizer. Ann Rev Cell Dev Biol 13:611–667

Heasman J (1997) Patterning the *Xenopus* blastula. Development 124:4179–4191

Hogan BLM (1996) Bone morphogenetic proteins: multifunctional regulators of vertebrate development. Genes Dev 10:1580–1594

Joly J-S, Joly C, Schulte-Merker S, Boulkebache H, Condamine H (1993) The ventral and posterior expression of the homeobox gene *eve1* is perturbed in dorsalized and mutant embryos. Development 119:1261–1275

Kane DA, Kimmel CB (1993) The zebrafish midblastula transition. Development 119:447–456

Kessler E, Takahara K, Biniminov L, Brusel M, Greenspan DS (1996) Bone morphogenetic protein-1: the type-1 procollagen C-proteinase. Science 271:360–362

Kimmel CB, Warga RM, Schilling TF (1990) Origin and organization of the zebrafish fate map. Development 108:581–594

Kishimoto Y, Lee K-H, Zon L, Hammerschmidt M, Schulte-Merker S (1997) The molecular nature of *swirl*: BMP2 function is essential during early dorsoventral patterning. Development 124:4457–4466

Kretschmar M, Massagué J (1998) SMADs: mediators and regulators of TGF-β signaling. Curr Opin Genet Dev 8:103–111

Luo G, Hofman C, Bronckers ALJJ, Sohocki M, Bradley A, Karsenty G (1995) BMP-7 is an inducer of nephrogenesis, and is also required for eye development and skeletal patterning. Genes Dev 9:2808–2820

Martínez-Barberá JP, Toresso H, Da Rocha S, Krauss S (1997) Cloning and expression of three members of the zebrafish Bmp family: *Bmp2a, Bmp2b* and *Bmp4*. Gene 198:53–59

Miller-Bertoglio VE, Fisher S, Sánchez A, Mullins MC, Halpern ME (1997) Differential regulation of *chordin* expression domains in mutant zebrafish. Dev Biol 192:537–550

Mullins MC, Hammerschmidt M, Haffter P, Nüsslein-Volhard C (1994) Large-scale mutagenesis in the zebrafish: in search of genes controlling development in a vertebrate. Curr Biol 4:189–202

Mullins MC, Hammerschmidt M, Kane DA, Odenthal J, Brand M, Eeden van FJM, Furutani-Seiki M, Granato M, Haffter P, Heisenberg C-P, Jiang Y-J, Kelsh RN, Nüsslein-Volhard C (1996) Genes establishing dorsoventral pattern formation in the zebrafish embryo: the ventral specifying genes. Development 123:81–93

Nikaido M, Tada M, Saij T, Ueno N (1997) Conservation of BMP signaling in zebrafish mesoderm patterning. Mech Dev 61:75–88

Piccolo S, Agius E, Lu B, Goodman S, Dale L, DeRobertis EM (1997) Cleavage of Chordin by Xolloid metalloprotease suggests a role for proteolytic processing in the regulation of Spemann organizer activity. Cell 91:407–416

Piccolo S, Y., Sasai Y, Lu B, De Robertis EM (1996) A possible molecular mechanism for Spemann organizer function: inhibition of ventral signals by direct binding of Chordin to BMP-4. Cell 85:589–598

Schulte-Merker S, Lee LJ, McMahon AP, Hammerschmidt M (1997) The zebrafish organizer requires *chordino*. Nature 387:862–863

Solnica-Krezel L, Stemple DL, Mountcastle-Shah E, Rangini Z, Neuhauss SCF, Malicki J, Schier A, Stanier DYR, Zwartkruis F, Abdelilah S, Driever W (1996) Mutations affecting cell fates and cellular rearrangements during gastrulation in the zebrafish. Development 123:67–80

Thomsen GH (1997) Antagonism within and around the organizer: BMP inhibitors in vertebrate body patterning. Trends Genet 13:209–211

Warga RM, Kimmel CB (1990) Cell movements during epiboly and gastrulation in zebrafish. Development 108:569–80

Wozney JM, Rosen V, Celeste AJ, Mitsock LM, Whiters MJ, Kris RW, Hewick RM, Wang EA (1988) Novel regulators of bone formation: molecular clones and activities. Science 242:1528–1534

Zhang H, Bradley A (1996) Mice deficient for BMP2 are nonviable and have defects in amnion/chorion and cardiac development. Development 122:2977–2986

Zhang J, Talbot WS, Schier AF (1998) Positional cloning identifies zebrafish one-eyed-pinhead as a permissive EGF-related ligand required during gastrulation. Cell 92:241–251

Zimmerman LB, Harland RM (1996) Bmp-4 function is blocked by high affinity binding to the Spemann organizer signal Noggin. Cell 85:599–606

7 Genetic Dissection of Heart Development

J.-N. Chen and M. C. Fishman

7.1 Introduction

Genetic analysis has been a powerful tool to identify genes and pathways in early patterning and cell fate determination in *Drosophila* and *Caenorhabditis elegans* (Nusslein-Volhard 1994). Lacking a vertebrate model system for large-scale genome-wide screens has limited a direct genetic approach toward vertebrate pattern formation in the past. Luckily, gene functions and pathways are often conserved from invertebrates to vertebrates and studying these pathways has been fruitful in understanding vertebrate pattern formation. Some of the best known examples include the role of the hox genes in patterning the vertebrate nervous system, the TGF-β family and wnt signaling pathways in dorsoventral patterning, and the hedgehog pathway in nervous system and limb patterning [for review see Pattern Formation during Development (1997) Cold Spring Harbor Symposium Quantum Biology, vol. 62].

Zebrafish has recently attracted significant attention from vertebrate developmental biologists, because it is amenable to both embryological and genetics studies. DNA/RNA injections, cell transplantation, in vivo time-lapse analysis, and lineage-tracing techniques are routinely used to analyze gene function and developmental processes in the zebrafish. In general, each pair of fish produces hundreds of progeny every week. The embryo is fertilized outside of the body and is transparent. Therefore, developmental processes can be followed from the single-cell stage to the time when organs are fully formed and functional. This makes zebrafish uniquely suitable for large-scale vertebrate genetic screens. Two large-scale genetic screens have been performed and more than two thousand mutations affecting specific developmental processes have been identified (Driever et al. 1996; Haffter et al. 1996). Recent progress in assembling genomic resources facilitates the cloning of these mutations.

Zebrafish is especially suited to studies of the embryonic heart. A functional heart is present in a prominent ventral location in the zebrafish embryo 24 hours after fertilization. The fish heart, like the heart of other vertebrates, consists of two major cell types, myocardium and endocardium, derived from precursors, bilateral mesodermal primordia. Soon after fusion, a border forms between what will become the atrium and the ventricle. In mammals and birds, septae later grow to subdivide both the embryonic atrium and the embryonic ventricle, generating a four-chambered heart. The ventricle later bends towards the right side of the embryo in the zebrafish starting at 33 hours after fertilization. This process is referred to as cardiac looping, and occurs in the heart of all vertebrates. After 2 days of development in the zebrafish, the ventricular myocardial layer differentiates into a thick-walled chamber, and is capable of generating systemic blood pressure. The valves start to differentiate at the borders of the chambers after 3 days of development to prevent backflow of blood (Chen and Fishman 1997). We have discovered in two large-scale genetic screens, mutations which affect each of the steps of heart morphogenesis, as well as its functional development of rate, rhythm, conduction, and contractility (Chen et al. 1996; Stainier et al. 1996).

The application of the zebrafish system to dissect pathways of functional development of the heart is discussed elsewhere (Warren and Fishman 1998). In this review, we focus on three issues of heart formation. At the cellular level, we focus on tracking the localization of cardiac precursors at various developmental stages and identifying the tissues and factors affecting cardiac cell fate determination. At the organ level, we focus on cardiac chamber formation. At the embryo level, we will discuss the embryonic left-right influence on the heart. Finally, we will discuss the prospects for applying the zebrafish system to studying organogenesis.

7.2 Heart Field

Although mesodermal in origin, myocyte migration paths differ among species. For example, in *Drosophila*, the mesodermal precursors, including the heart, are originated ventrally and later migrate and become specified at dorsal sites (Azpiazu and Frasch 1993; Bodmer 1993). In *Xenopus*, the heart precursors originate from the dorsolateral tissues and migrate ventrally (for review see Fishman and Chien 1997). Prior to formation of the primitive heart mesoderm, myocytes reside lateral to the midline from where they fuse (at the midline) and form the heart.

In the zebrafish, the location and the migratory path of the cardiac precursors have been traced embryologically by single cell injection lineage-tracing technique (Stainier et al. 1993). The cardiac precursors are localized at the margin of the ventral hemisphere of the zebrafish embryo as early as the blastula stage. The most ventrally localized cells have the highest probability of contributing to the heart, a property which diminishes toward the dorsal side. Some of the progeny of the ventral blastomeres involute during gastrulation, migrate toward the embryonic midline in the lateral plate and eventually populate the heart (Stainier et al. 1993).

Nkx2.5 is a vertebrate homologue of the *Drosophila* homeodomain gene *tinman*. Mutation in *tinman* results in defective heart formation (Bodmer 1993). In all vertebrates analyzed including, fish, frog, chick, and mouse, *Nkx2.5* is expressed in a region which at least overlaps with the position of lineage-derived cardiac precursors (for review see Harvey 1996). Overexpression of *Nkx2.5* causes an enlarged heart in the

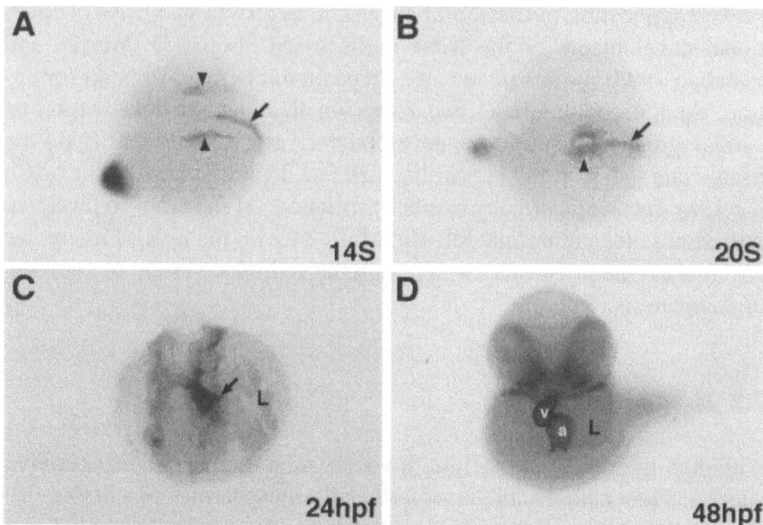

Fig. 1A–D. Zebrafish heart development. The cardiac precursors originate from the ventral hemisphere of the zebrafish embryo. At the onset of gastrulation these cells involute and migrate toward the embryonic midline. At the somitogenesis stage these cells reside on either side of embryo in the lateral plate (**A**). By the 20-somite stage, the bilateral cardiac primordia fuse at the midline (**B**) and form the primitive heart tube. **C** At 24 h after fertilization, the atrial end of the primitive heart tube moves to the left side (*L*) of the zebrafish embryo, termed cardiac jogging. The left-jog heart will then gradually swing back to the midline. **D** By 48 h after fertilization, the ventricle of the midline heart bends to the right of the atrium, termed cardiac looping. The heart is labeled by *Nkx2.5* expression in **C** (*arrow*), **A,** and **B** (*arrowhead*), and MF20 in **D**. The notochord is labeled by the expression of *ntl* (*arrow*) in **A** and **B**

frog and zebrafish (Chen and Fishman 1996; Cleaver et al. 1996). A high level of *Nkx2.5* gene activity can induce low-level cardiac gene expression at ectopic locations in the zebrafish embryos and cultured fibroblasts (Chen and Fishman 1996). In contrast to the "no heart" phenotype in *Drosophila*, targeted gene disruption of *Nkx2.5* in the mouse arrests heart development by the looping stage (Lints et al. 1993). This partial disruption of the cardiogenic pathway is likely due to gene redundancy. *XNkx2.3* and *XNkx2.5* are both *Xenopus* homologues of *tinman*, and are both expressed in the cardiogenic region (Evans et al.

1995). When the activity of either *XNkx2.3* or *XNkx2.5* alone is blocked in *Xenopus* embryos, the heart is mildly affected. However, heart formation is abolished when both *Nkx2.3* and *Nkx2.5* gene activities are blocked (Fu et al. 1998; Grow and Krieg 1998).

As shown by normal *Nkx2.5* expression in some non-cardiogenic regions of many species, and the inability of overexpression to cause complete myogenesis, *Nkx2.5* is insufficient to cause cardiomyocyte differentiation. One important component may be the persistence past a certain time of development. In *Drosophila*, of the many *tinman*-expressing mesoderm cells, it is only cells which maintain *tinman* gene activity which adopt the cardiac fate (Frasch 1995). The expression of *dpp* in the adjacent ectoderm is required for maintaining *tinman* gene expression. Disruption of *dpp* activity abolishes *tinman* gene expression in the cardioblasts, and therefore abolishes heart formation (Frasch 1995). Unlike other vertebrates studied to date, in the zebrafish, the *Nkx2.5* expression pattern corresponds to the position of cardiogenic precursors as early as the onset of gastrulation (Chen and Fishman 1996). At the somitogenesis stage, the *Nkx2.5* expression is restricted to the lateral plate (Fig. 1). At the time as the bilateral cardiac primordia fuse at the midline, the lateral mesodermal cells posterior to the notochord turn off *Nkx2.5* expression. As fate-mapped by a laser-based technique, these cells do not populate the heart. Only the lateral mesodermal cells anterior to the notochord maintain *Nkx2.5* gene activity and populate the heart (Serbedzija et al. 1998).

Classic embryological experiments showed that at early developmental stages, the embryos can compensate for the loss of organ primordia (Copenhaver 1924). In the zebrafish, the ability to compensate for loss of cardiac precursors continues until the heart fuses at the midline (Serbedzija et al. 1998). Interestingly, the cells transiently expressing *Nkx2.5* adjacent to the notochord do not appear to contribute to this regulatory compensation (Serbedzija et al. 1998). Together with the fate mapping analysis, this implies that the notochord has a negative effect on heart development. In fact, after the ablation of the tip of the notochord, the adjacent, normally non-cardiac *Nkx2.5* cells do develop the ability to populate the heart (Goldstein and Fishman 1998).

How does the notochord regulate heart formation? One possibility is through regulating *Nkx2.5* gene activity via the *BMP* pathway. In the notochord-ablated zebrafish embryos, the *Nkx2.5* expression domain extends more posteriorly (Goldstein and Fishman 1998). In *Drosophila*, *tinman* expression in the cardiac progenitors is dependent on *dpp* activity from the adjacent ectoderm (Frasch 1995). A DNA-binding site of *medea*, a downstream responsive gene of *dpp*, is present in the *tinman* promoter and is required for proper *tinman* gene expression (Xu et al. 1998). The vertebrate homologue of *dpp*, *BMP4*, is expressed in the zebrafish heart and *BMP4* antagonists, such as *noggin* and *chordin,* are expressed in the notochord (Bauer et al. 1998), implying that the regulatory circuits of *dpp-tinman* may be conserved in vertebrates.

7.3 Chamber Formation

A multi-chambered heart is a feature of all vertebrates, but not of primitive chordates, suggesting that it is a recent evolutionary innovation of vertebrates. It serves to ensure unidirectional flow and high-pressure circulation (Fishman and Chien 1997), in turn permitting significant increase of the size. How chambers becomes demarcated, how the chamber fates are specified, and how the chamber-specific cell types differentiate is poorly understood.

The helix-loop-helix gene, *dHAND*, is the only one known to be important in chamber formation. It is expressed initially in the cardiogenic region of the developing mouse heart and later restricted to the right ventricle and some neural crest derivatives (Srivastava et al. 1995; Srivastava et al. 1997). Targeted gene disruption of *dHAND* abolishes the right ventricle (Srivastava et al. 1997). Mutation in *dHAND* also abolishes cardiac ventricle formation in the zebrafish embryos (Stainier, personal communication). Furthermore, a downstream factor of *dHAND*, *ufd1* gene, is deleted in human patients with cardiac and craniofacial anomalies (Yamagishi et al. 1999). How genes in the dHAND pathway control cardiac chamber formation is not known, but evidence suggests that *dHAND* is required for cardiac cell survival (Srivastava 1999). Other pathways or genes might be involved in cardiac chamber formation. Mutant embryos of *pandora* and *lonely atrium* do not have a ventricular structure (Fig. 2) (Chen et al. 1996; Stainier et al. 1996). Molecular cloning of these genes will shed light on their regulation of cardiac chamber formation.

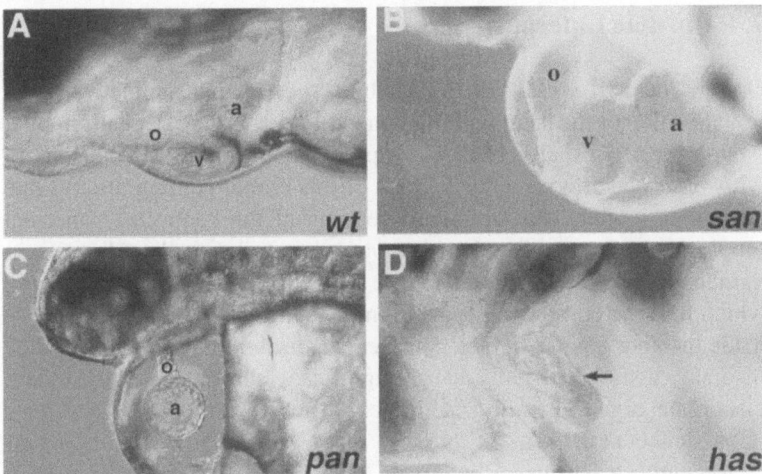

Fig. 2A–D. Zebrafish cardiac chamber mutations. **A** After 3 days of develop-ment, the zebrafish heart assumes a prominent ventral position and the atrium (*a*), ventricle (*v*), and outflow tract (*o*) are well developed. Three classes of mutations affecting cardiac chamber formation have been isolated from the ze-brafish genetic screens. **B** Mutant embryos of *santa* (*san*) develop an enlarged heart. **C** *pandora* (*pan*) abolishes the ventricle formation, and (**D**) the ventricu-lar cells are encompassed within the atrium in the *heart and soul* (*has*) mutant embryos

.Another important step in chamber generation is their patterning and orientation. Several genes are important to this process, as shown by the large-scale genetic screens in zebrafish (Fig. 2) (Stainier et al. 1996). In mutant embryos of *heart and soul*, the ventricular cells appear to be encompassed within the atrium, suggesting a role in establishing the proper positional relationship between the chambers. The genes *santa*, *heart of glass,* and *valentine* are involved in concentric growth of myo-cardium of the ventricle. Normally, the myocardial cells undergo divi-sion and growth in the ventricle, generating the normal thick-walled chamber. In these mutant embryos, the ventricular wall remains thin and the heart dilates.

7.4 Cardiac Laterality

Organs must fit within an overall body plan. For example, physical juxtaposition of the heart with other neighboring organs depends upon it having a certain shape, as does its proper and functional connectivity with veins and arteries. One element of this proper configuration is due to cardiac looping, the rightward bending of the ventricle. Abnormal looping is often associated with severe congenital cardiac diseases in humans (for review see Goldstein et al. 1998). It is not simply looping which must be correct, but looping in the context of laterality decisions made by other organs. In fact, complete reversal of organ positions, situs inversus, does not usually cause significant abnormalities. However, discordant organ laterality, heterotaxy, often results in severe defects (Goldstein et al. 1998).

Recently, a few genes involved in left-right patterning have been identified. Genetic or embryological manipulation of such gene activities often leads to randomization of cardiac looping (for review see Ramsdell and Yost 1998). Interestingly, with few exceptions, most of the molecules are not asymmetrically expressed in the organs themselves. It is quite possible that these genes provide embryonic signals which guide organ laterality, but are not the responding molecules within the organs. Which genes does the heart use to interpret or respond to the embryonic left–right signals? There are two good candidates for such a role. The gene *pitx2* is a vertebrate *bicoid* homologue, which is asymmetrically expressed in the heart and the gut in mouse, chicken, and frog embryos. Experimental manipulation of *pitx2* gene activity in the frog and chick embryos causes randomization of organ laterality (Logan et al. 1998; Piedra et al. 1998; Ryan et al. 1998; Yoshioka et al. 1998). *BMP4* is asymmetrically expressed in the zebrafish primitive heart tube, soon after the bilateral primordia fuse at the midline. This asymmetry is disrupted in the zebrafish laterality mutants (see below). Furthermore, experimental disruption of *BMP4* signaling in the heart results in randomization of cardiac looping (Chen et al. 1997).

A genome-wide survey for genes establishing organ laterality is of interest as the first step toward genetic dissection of this pathway. Cardiac looping has classically been utilized as the first evidence of left-right asymmetry in vertebrates. However, it is difficult to screen for cardiac laterality mutations using cardiac looping as the only assay. A proper looping is dependent on factors in addition to laterality decisions, for example, proper cardiac function. Therefore, it is difficult to interpret the cause of mutant phenotypes. Interestingly, in the zebrafish, 12 hours before looping, the arterial end of the heart transiently moves to the left side of the embryo. This process is referred to as cardiac jogging (Fig. 1) (Chen et al. 1997). Although jogging has not been described in other vertebrates, it may not be a zebrafish-specific feature. In fact, the mouse embryonic heart also shifts to the left prior to looping (En Li and Richard Harvey, personal communication). Jogging appears to be solely affected by laterality decisions, therefore, it is a more reliable indicator for cardiac laterality. Using both jogging and looping as morphological assays for left-right patterning, 279 mutations from the Tübingen zebrafish stock center were screened and 21 were found to have cardiac laterality defects (Chen et al. 1997).

These mutations confirm *BMP4* as a mediator between the embryonic left-right signals and the cardiac laterality. Normally, *BMP4* expression is left-predominant in the heart soon after the bilateral primordia fuse at the midline. This asymmetric pattern is randomized in the mutant embryos with an abnormal jog. A left-predominant *BMP4* expression correlates with a left-jog. Right-predominant *BMP4* expression correlates with a right-jog. A symmetric *BMP4* expression pattern correlates with randomized jogging (Chen et al. 1997).

These mutants also reveal jogging as a reliable indicator for embryonic laterality. The direction of jogging reflects the preceding *BMP4* asymmetry and is predictive of the direction of the subsequent looping. In both wild-type and mutant embryos, a left-jogged heart always loops to the right. A right-jogged heart always loops to the left. If the heart fails to jog and stays at the midline, the subsequent looping is randomized, and could be to the right or to the left, or the heart remains unlooped (Chen et al. 1997).

The cardiac laterality screen was based on the existing collection of zebrafish mutants. Therefore, all the cardiac laterality mutants identified in this screen bear additional defects. Based on the phenotypes, these mutants can be categorized into three classes: those with dorsoventral defects, those with midline defect, and those with a curved body shape (Chen et al. 1997). This supports the notion that the proper patterning of the left-right axis is dependent on the proper patterning of the dorsoventral axis and the midline structures (Danos and Yost 1995; Danos and

Yost 1996). Molecular cloning of these mutations are underway in the hope that they will provide further insight into the signaling between the embryonic axis and the organs.

7.5 Organogenesis Screen

Genetics has been a powerful tool to identify genes and pathways. Currently many attributes of vertebrate development are extrapolated from *Drosophila*. However, this approach is limited when exploring vertebrate organogenesis, because invertebrate and vertebrate organs are fundamentally different. For example, *Drosophila* has an open circulatory system, while vertebrates develop a seamless and closed vasculature. The *Drosophila* heart is a muscular tube consisting of one major cell type, the cardiomyocyte. The vertebrate heart consists of multiple chambers and two major cell types, the inner endocardial cells and the outer myocardial cells, which also differ between chambers. Therefore, a vertebrate genome-wide screen focusing on organogenesis is essential to help understand the formation and function of vertebrate organs.

Nearly 2000 mutations affecting different aspects of zebrafish development have been isolated in two large-scale zebrafish genetic screens, at Massachusetts General Hospital and at the Max-Plank Institute for Developmental Biology in Tübingen (Driever et al. 1996; Haffter et al. 1996). Relatively low numbers of organogenesis mutations were identified in these screens, with the exception of the heart mutants. One likely explanation is that the heart is placed at a prominent ventral position, but other organs are placed in a rather deeper layer of the zebrafish embryo. Therefore, it is more difficult to visualize the visceral organs. The phenotypes of these mutations can be quite subtle and are easily overlooked. Furthermore, organ development is a relatively late event in embryogenesis. If genes required for organ development are also required for early events, the organotypic phenotypes will be masked by early phenotypes. Therefore, it is likely that a simple visual screen is not optimal when searching for such mutations and specific assays might be necessary to reveal these mutations. This has proven to be the case by a recent pilot screen focusing on organogenesis. The design of the pilot screen is the classic F3 screen for recessive mutations (Haffter et al. 1996). A total of 750 mutagenized genomes were screened using specific in situ and antibody probes, in combination with the classic visual analysis, searching for organogenesis mutations. We discovered mutations specifically affecting heart formation and/or function, organ laterality, the vascular system or the development of kidney, and the gut and pancreas (Chen and Fishman, unpublished data). Detail characterization and molecular cloning of these mutations is ongoing.

7.6 Cloning Zebrafish Mutations

Genetic screens provide means to dissect genetic pathways. The cloning of these mutations leads to molecular understanding of the pathways. A small fraction of the zebrafish mutations are generated by retrovirus-based insertion mutagenesis (Gaiano et al. 1996). The advantage of this approach is that the mutation is "tagged" by the retrovirus and is easy to clone. Once a desirable mutant phenotype is identified, the molecular nature of the mutated gene can easily be revealed. However, the efficiency of the retrovirus-based mutagenesis is still significantly lower than chemical mutagenesis and is labor-intensive (Schier et al. 1996).

Most of the zebrafish mutations currently available are chemically-induced point mutations. Although it is easier to generate mutations with this approach, it is much more difficult to clone them. Up to now, only two zebrafish mutations have been reported to be cloned positionally (Brownlie et al. 1998; Zhang et al. 1998). However, zebrafish genomic resources have grown rapidly in the past few years, facilitating positional cloning of the mutations. The first step toward cloning is to place these mutations on a genetic map. Currently, about 2400 markers are placed on the SSLP-based genetic map, giving the resolution of 0.9 cM on average (Shimoda et al. 1999). This resolution allows one to initiate physical mapping and walking. Many mutations have been successfully placed on the SSLP map.

The most expeditious cloning strategy is through candidate genes. About 20 zebrafish mutations have been cloned using as candidates, genes identified because of their expression pattern and/or map position (for examples see Hild et al. 1999; Kishimoto et al. 1997; Rauch et al. 1997). Mapping cDNA or EST by polymorphism (Gates et al. 1999) can be quite time-consuming, but is rapid when using radiation hybrid panels. Two zebrafish radiation hybrid panels are now available (Geisler et al. 1999; Hukriede et al. 1999). Hundreds of SSLP markers have been typed on both panels as anchors. Mapping large number of ESTs and cDNAs on the RH panel will accelerate the molecular analysis of mutations by providing large pool of candidates. Furthermore, mapping large number of genes will provide the means to compare syntenic relationships between species.

7.7 Conclusion

In the past, molecular analysis of organ formation was limited to the study of known genes or pathways, but an application of the logical steps of organogenesis cannot be achieved so. Zebrafish offer a new opportunity to apply genetics in combination with embryology. The first screens show that single steps can be revealed genetically for organ formation, in a manner akin to those for body formation in *Drosophila*. For example, the genetic steps for chamber formation and patterning, and for left-right laterality are made evident by mutational analysis. Each such gene provides a molecular handle to pathways, other components of which may be discovered by biochemical or genetic techniques. Other organs, including the kidney, gut and pancreas, are proving receptive to this approach as well.

References

Azpiazu N, Frasch M (1993) tinman and bagpipe: two homeo box genes that determine cell fates in the dorsal mesoderm of *Drosophila*. Genes Dev 7:1325–1340

Bauer H, Meier A, Hild M, Stachel S, Economides A, Hazelett D, Harland RM, Hammerschmidt M (1998) Follistatin and noggin are excluded from the zebrafish organizer. Dev Biol 204:488–507

Bodmer R (1993) The gene tinman is required for specification of the heart and visceral muscles in *Drosophila*. Development 118:719–729

Brownlie A, Donovan A, Pratt SJ, Paw BH, Oates AC, Brugnara C, Witkowska HE, Sassa S, Zon LI (1998) Positional cloning of the zebrafish sauternes gene: a model for congenital sideroblastic anaemia. Nat Genet 20:244–250

Chen J-N, Fishman MC (1997) Development of Cardiovascular System: Molecules to Organisms. Cambridge University Press, Cambridge

Chen J-N, Fishman MC (1996) Zebrafish tinman homolog demarcates the heart field and initiates myocardial differentiation. Development 122:3809–3816

Chen J-N, Haffter P, Odenthal J, Vogelsang E, Brand M, van Eeden FJ, Furutani-Seiki M, Granato M, Hammerschmidt M, Heisenberg CP, Jiang YJ, Kane DA, Kelsh RN, Mullins MC, Nüsslein-Volhard C (1996) Mutations affecting the cardiovascular system and other internal organs in zebrafish. Development 123:293–302

Chen J-N, van Eeden FJ, Warren KS, Chin A, Nüsslein-Volhard C, Haffter P, Fishman MC (1997) Left-right pattern of cardiac BMP4 may drive asymmetry of the heart in zebrafish. Development 124:4373–4382.

Cleaver OB, Patterson KD, Krieg PA (1996) Overexpression of the tinman-related genes XNkx-2.5 and XNkx-2.3 in *Xenopus* embryos results in myocardial hyperplasia. Development 122:3549–3556

Copenhaver WM (1924) Experiments on the development of the heart of *Amblystoma punctatum*. J Exp Zool 43:321–371

Danos MC, Yost HJ (1995) Linkage of cardiac left-right asymmetry and dorsal-anterior development in *Xenopus*. Development 121:1467–1474

Danos MC, Yost HJ (1996) Role of notochord in specification of cardiac left-right orientation in zebrafish and *Xenopus*. Dev Biol 177:96–103

Driever W, Solnica-Krezel L, Schier AF, Neuhauss SC, Malicki J, Stemple DL, Stainier DY, Zwartkruis F, Abdelilah S, Rangini Z, Belak J, Boggs C (1996) A genetic screen for mutations affecting embryogenesis in zebrafish. Development 123:37–46

Evans SM, Yan W, Murillo MP, Ponce J, Papalopulu N (1995) tinman, a *Drosophila* homeobox gene required for heart and visceral mesoderm specification, may be represented by a family of genes in vertebrates: XNkx-2.3, a second vertebrate homologue of tinman. Development 121:3889–3899

Fishman MC, Chien KR (1997) Fashioning the vertebrate heart: earliest embryonic decisions. Development 124:2099–2117

Frasch M (1995) Induction of visceral and cardiac mesoderm by ectodermal Dpp in the early *Drosophila* embryo. Nature 374:464–467

Fu Y, Yan W, Mohun TJ, Evans SM (1998) Vertebrate tinman homologues XNkx2–3 and XNkx2–5 are required for heart formation in a functionally redundant manner. Development 125:4439–4449

Gaiano N, Amsterdam A, Kawakami K, Allende M, Becker T, Hopkins N (1996) Insertional mutagenesis and rapid cloning of essential genes in zebrafish. Nature 383:829–832

Gates MA, Kim L, Egan ES, Cardozo T, Sirotkin HI, Dougan ST, Lashkari D, Abagyan R, Schier AF, Talbot WS (1999) A genetic linkage map for zebrafish: comparative analysis and localization of genes and expressed sequences. Genome Res 9:334–347

Geisler R, Rauch GJ, Baier H, van Bebber F, Brobeta L, Dekens MP, Finger K, Fricke C, Gates MA, Geiger H, Geiger-Rudolph S, Gilmour D, Glaser S, Gnugge L, Habeck H, Hingst K, Holley S, Keenan J, Kirn A, Knaut H, Lashkari D, Maderspacher F, Martyn U, Neuhauss S, Neumann C, Nicolson T, Pelegri F, Ray R, Rick JM, Roehl H, Roeser T, Schauerte HE, Schier AF, Schonberger U, Schonthaler HB, Schulte-Merker S, Seydler C, Talbot WS, Weiler C, Nüsslein-Volhard C, Haffter P (1999) A radiation hybrid map of the zebrafish genome. Nat Genet 23:86–89

Goldstein AM, Fishman MC (1998) Notochord regulates cardiac lineage in zebrafish embryos. Dev Biol 201:247–252

Goldstein AM, Ticho BS, Fishman MC (1998) Patterning the heart's left-right axis: from zebrafish to man. Dev Genet 22:278–287

Grow MW, Krieg PA (1998) Tinman function is essential for vertebrate heart development: elimination of cardiac differentiation by dominant inhibitory mutants of the tinman-related genes, XNkx2–3 and XNkx2–5. Dev Biol 204:187–196

Haffter P, Granato M, Brand M, Mullins MC, Hammerschmidt M, Kane DA, Odenthal J, van Eeden FJ, Jiang YJ, Heisenberg CP, Kelsh RN, Furutani-Seiki M, Vogelsang E, Beuchle D, Schach U, Fabian C, Nüsslein-Volhard C (1996) The identification of genes with unique and essential functions in the development of the zebrafish, *Danio rerio*. Development 123:1–36

Harvey RP (1996) NK-2 homeobox genes and heart development. Dev Biol 178:203–216

Hild M, Dick A, Rauch GJ, Meier A, Bouwmeester T, Haffter P, Hammerschmidt M (1999) The smad5 mutation somitabun blocks Bmp2b signaling during early dorsoventral patterning of the zebrafish embryo. Development 126:2149–2159

Hukriede NA, Joly L, Tsang M, Miles J, Tellis P, Epstein JA, Barbazuk WB, Li FN, Paw B, Postlethwait JH, Hudson TJ, Zon LI, McPherson JD, Chevrette M, Dawid IB, Johnson SL, Ekker M (1999) Radiation hybrid mapping of the zebrafish genome [In Process Citation]. Proc Natl Acad Sci USA 96:9745–9750

Kishimoto Y, Lee KH, Zon L, Hammerschmidt M, Schulte-Merker S (1997) The molecular nature of zebrafish swirl: BMP2 function is essential during early dorsoventral patterning. Development 124:4457–4466

Lints TJ, Parsons LM, Hartley L, Lyons I, Harvey RP (1993) Nkx-2.5: a novel murine homeobox gene expressed in early heart progenitor cells and their myogenic descendants. Development 119:969

Logan M, Pagan-Westphal SM, Smith DM, Paganessi L, Tabin CJ (1998) The transcription factor Pitx2 mediates situs-specific morphogenesis in response to left-right asymmetric signals. Cell 94:307–317

Nüsslein-Volhard C (1994) Of flies and fishes. Science 266:572–574

Piedra ME, Icardo JM, Albajar M, Rodriguez-Rey JC, Ros MA (1998) Pitx2 participates in the late phase of the pathway controlling left-right asymmetry. Cell 94:319–324

Ramsdell AF, Yost HJ (1998) Molecular mechanisms of vertebrate left-right development. Trends Genet 14:459–465

Rauch GJ, Hammerschmidt M, Blader P, Schauerte HE, Strahle U, Ingham PW, McMahon AP, Haffter P (1997) Wnt5 is required for tail formation in the zebrafish embryo. Cold Spring Harb Symp Quant Biol 62:227–234

Ryan AK, Blumberg B, Rodriguez-Esteban C, Yonei-Tamura S, Tamura K, Tsukui T, de la Pena J, Sabbagh W, Greenwald J, Choe S, Norris DP, Robertson EJ, Evans RM, Rosenfeld MG, Izpisua Belmonte JC (1998) Pitx2 determines left-right asymmetry of internal organs in vertebrates. Nature 394:545–551

Schier AF, Joyner AL, Lehmann R, Talbot WS (1996) From screens to genes: prospects for insertional mutagenesis in zebrafish. Genes Dev 10:3077–3080

Serbedzija GN, Chen J-N, Fishman MC (1998) Regulation in the heart field of zebrafish. Development 125:1095–1101

Shimoda N, Knapik EW, Ziniti J, Sim C, Yamada E, Kaplan S, Jackson D, de Sauvage F, Jacob H, Fishman MC (1999) Zebrafish genetic map with 2000 microsatellite markers. Genomics 58:219–232

Srivastava D (1999) HAND proteins: molecular mediators of cardiac development and congenital heart disease. Trends Cardiovasc Med 9:11–8

Srivastava D, Cserjesi P, Olson EN (1995) A subclass of bHLH proteins required for cardiac morphogenesis. Science 270:1995–1999

Srivastava D, Thomas T, Lin Q, Kirby ML, Brown D, Olson EN (1997) Regulation of cardiac mesodermal and neural crest development by the bHLH transcription factor, dHAND. Nat Genet 16:154–160

Stainier DY, Fouquet B, Chen JN, Warren KS, Weinstein BM, Meiler SE, Mohideen MA, Neuhauss SC, Solnica-Krezel L, Schier AF, Zwartkruis F, Stemple DL, Malicki J, Driever W, Fishman MC (1996) Mutations affecting the formation and function of the cardiovascular system in the zebrafish embryo. Development 123:285–292

Stainier DY, Lee RK, Fishman MC (1993) Cardiovascular development in the zebrafish. I. Myocardial fate map and heart tube formation. Development 119:31–40

Warren KS, Fishman MC (1998) "Physiological genomics": mutant screens in zebrafish. Am J Physiol 275:H1–7

Xu X, Yin Z, Hudson JB, Ferguson EL, Frasch M (1998) Smad proteins act in combination with synergistic and antagonistic regulators to target Dpp responses to the *Drosophila* mesoderm. Genes Dev 12:2354–2370

Yamagishi H, Garg V, Matsuoka R, Thomas T, Srivastava D (1999) A molecular pathway revealing a genetic basis for human cardiac and craniofacial defects. Science 283:1158–1161

Yoshioka H, Meno C, Koshiba K, Sugihara M, Itoh H, Ishimaru Y, Inoue T, Ohuchi H, Semina EV, Murray JC, Hamada H, Noji S (1998) Pitx2, a bicoid-type homeobox gene, is involved in a lefty-signaling pathway in determination of left-right asymmetry. Cell 94:299–305

Zhang J, Talbot WS, Schier AF (1998) Positional cloning identifies zebrafish one-eyed pinhead as a permissive EGF-related ligand required during gastrulation. Cell 92:241–251

8 Eph Receptors and Ephrins Are Key Regulators of Morphogenesis

N. Holder, L. Durbin, and J. Cooke

8.1 Introduction

Morphogenesis is the creation of form during development. In all vertebrate embryos cells move and change shape in precise patterns at specific times to generate the tissues and layers of the forming body. The molecular basis of the control of morphogenesis is not well understood. The Eph receptors and their ligands the ephrins, represent a cellular signalling system that is important for the correct enactment of morphogenetic processes in the vertebrate and invertebrate embryo. These proteins are known to be key players in a number of morphogenetic events in the vertebrate embryo including axon guidance in the developing nervous system, neural crest migration and segmentation of the paraxial mesoderm and the hindbrain region of the neural plate. This review outlines the evidence for these conclusions and focuses mainly, but not exclusively, on the role of Eph/ephrin signalling in the nervous system.

Receptor tyrosine kinases (RTKs) are membrane-spanning proteins with an extracellular ligand-binding domain and an intracellular kinase domain. There are at least 14 subfamilies of RTKs (Van der Geer et al. 1994), and the Eph subfamily is the largest (Orioli and Klein 1997; Pasquale 1997; Tuzi and Gullick 1994). Each subfamily has characteristic ligand-binding and kinase domains and is activated by a distinct ligand or group of ligands. Many RTKs play roles in a broad range of processes in development. In 1987, Hirai et al. described the cloning and characterisation of the first member of the Eph subfamily, EphA1 (Hirai et al. 1987). To date as many as 14 genes have been described which are related to Eph by sequence and by general characteristics of their kinase and extracellular domains. Members of this subfamily of receptors have been isolated and characterised in a range of vertebrate species, including human, mouse, rat, chicken, *Xenopus* and zebrafish. They have also been isolated in invertebrates (George et al. 1998), but the invertebrate genes have been studied much less intensively than their vertebrate counterparts.

Ligands for the Eph family of RTKs, now known as ephrins (Eph Nomenclature Committee 1997) were identified in 1994 (Bartley et al. 1994). The receptors are now termed EphA or B depending on the class of ephrins that they bind (Eph Nomenclature Committee 1997; Table 1). The ephrins (Pandey et al. 1995) can be grouped into two classes: those which are connected to the membrane by a GPI linkage, called ephrin-A proteins (which bind to EphA receptors) and those with membrane-spanning and intracellular domains, called ephrin-B proteins (which bind to EphB receptors). The only receptor able to bind class A and class B ephrins is EphA4 (Gale et al. 1996). Like the receptors, the ephrins are dynamically expressed in the vertebrate embryo in various regions and tissues in the forming mesoderm, endoderm and nervous system. It has recently been shown that the transmembrane ligands, ephrin-B1 and -B2, can themselves become phosphorylated on their intracellular domains following binding and activation by receptors in an adjacent cell (Bruckner et al. 1997; Holland et al. 1996). A further key feature of ligand function is that they need to be membrane-bound to efficiently cluster and activate receptor signalling (Davis et al. 1994). This means that soluble forms of the ligands are likely to act as dominant-negative

Table 1. Members of the Eph and ephrin families

New Generic	Old Human	Mouse	Rat	Chicken	*Xenopus*[a]	Zebrafish[b]
Receptors						
EphA1	Eph					
EphA2	Eck	Myk2, Sek2			G42/G50	rtk6
EphA3	Hek	Mek4	Tyro4	Cek4		rtk2
EphA4	Hek8	Sek1	Tyro1	Cek8	Pag	rtk1
EphA5	Hek7	Bsk	Ehk1, Rek7	Cek7		rtk7
EphA6			Ehk2			
EphA7	Hek11	Mdk1, Ebk	Ehk3			
EphA8	Eek	Ptk4	Eek			rtk4/ZDK
EphB1	Net		Elk	Cek6	Xelk	
EphB2	Erk, Hek5, Drt		Nuk, Sek3	Tyro5	Cek5	rtk3
EphB3	Hek2	Sek4, Mdk5	Tyro6	Cek10		
EphB4	Htk	Myk1, Mdk2				rtk5/8
EphB5				Cek9		
EphB6		Mep				
Ligands						
ephrin-A1	Lerk1, B61					L1
ephrin-A2		Elf1		Cek7-L		L3
ephrin-A3	Lerk3		Ehk1-L			
ephrin-A4	Lerk4					
ephrin-A5	Lerk7		All	Rags		L2/4
ephrin-B1	Lerk2		Elk-L	Cek5-L	XLerk	
ephrin-B2	Lerk5, Htk-L			Elf-2		L5
ephrin-B3	Nlerk2		Elk-L3			

Recognised nomenclature is on the left, original names for each member is given for reference for the particular species concerned. (Based on Orioli and Klein 1997).
[a]Orthologous relationships of several *Xenopus* receptor and ephrin genes are still to be resolved (Brandli and Kirschner 1995; Scales et al. 1995; Weinstein et al. 1996).
[b]Two *Xenopus* or zebrafish receptors or ephrins separated by a dash represent probable paralogues which exist due to genome duplication events.

proteins as they can bind to a receptor but fail to cluster and activate them.

The formation of cell processes is fundamental to the movement of cells and therefore to morphogenesis. Two well-studied examples of cell process formation and cell movement in embryos are growth cone extension from differentiating neurons and neural crest cell migration. We outline below the evidence that Eph/ephrin signalling is an important component of the mechanisms underlying both events.

8.2 Eph/Ephrin Signalling Functions in Axon Guidance in the Peripheral and Central Nervous System

A vital aspect of the neuronal differentiation process is the production of an axon which grows and seeks a specific synaptic partner. It is evident from work on the retinotectal system (see below) that Eph signalling is involved in guiding axon growth. Initial evidence for such a role in the formation of centrally and peripherally projecting axon tracts arose from studies of the EphB2 receptor (Pasquale et al. 1992). EphB2 is expressed in a number of domains in the chick and mouse brain, including the retina. It has been shown that EphB2 in the chick retina is highly phosphorylated, particularly during the phase when interneuronal contacts are established (Pasquale et al. 1994). To further implicate EphB2 function in the formation of appropriate neuronal contacts, this receptor has been immunolocalised to the surface of growth cones of spinal motor neuron and oculomotor neuron axons from the onset of their growth towards their respective targets (Henkemeyer et al. 1994). Furthermore, functional studies using a different receptor, EphA5, and a ligand by which it is activated, ephrin-A5, showed directly that Eph receptor signalling is involved in axon fasciculation (Winslow et al. 1995). In these experiments fasciculation of axons from cortical neurons growing on astrocytes was inhibited by soluble forms of both the receptor and the ligand which are assumed to act in a dominant-negative manner. The strongest evidence to date implicating EphB2 in axon outgrowth in the embryo comes from targeted mutation studies in the mouse (Henkemeyer et al. 1996). An embryo homozygous for loss of EphB2 function lacks part of the anterior commissure. However, as the neurons projecting axons across the anterior commissure do not express EphB2, the authors suggest that the phenotype may reflect loss of signalling through the ligand to which it normally binds. Loss of commissural axons is more dramatic in mice null for both EphB2 and EphB3 (Orioli et al. 1996). In such animals the anterior commissure and the corpus callosum are affected as well as the forming palate, an area of the embryo in which both receptors are normally expressed. It was further observed that at least one CNS axon bundle running in the anterior-posterior direction, the habenular-interpeduncle tract, was partially defasciculated in the double mutants, although the projection to its target, the ventral midbrain, appeared normal.

The identification of two ephrins showing a graded distribution from posterior to anterior in the developing chick and mouse tectum suggested a role for Eph signalling in establishing appropriate connections in the retinotectal system. Drescher et al. purified an Eph ligand, ephrin-A5, from the chick tectum, based on its expression in the tectum and the fact that it is a GPI-linked protein (Drescher et al. 1995). Bioassays, devised to characterise the growth of chick retinal axons over the tectum, showed that membranes isolated from tectal cells possessed a collapsing activity for growth cones from temporal but not nasal retinal ganglion cells. Ephrin-A5 mimicked this collapsing activity. The existence of a second ligand in the tectum, ephrin-A2, was shown in the embryo by binding of a chimaeric protein in which the extracellular domain of a receptor was linked to alkaline phosphatase. The ephrin-A2 cDNA was then cloned using an expression library (Cheng and Flanagan 1994).

In the chick the ephrin-A5 and ephrin-A2 ligands are assumed to interact with two Eph receptors: EphA4, which is expressed uniformly across the retina, and EphA3, whose expression is graded across the retina, with a high point in the temporal region (Cheng et al. 1995). In the tectum, the expression domains of the ligands differ from each other, with ephrin-A2 extending more anteriorly than ephrin-A5. With variations in ligand-binding specificities and the graded distributions of both the ligands and receptors (Monschau et al. 1997), it is feasible that sufficient information can be provided by Eph signalling to resolve the retinotopic map. In addition, Eph receptors and ligands are spatially regulated with respect to the dorsal-ventral axis of the eye. For instance, EphB2 is expressed more strongly in the ventral than dorsal retina (Holash and Pasquale 1995; Kenny et al. 1995). Similarly, ephrins of the A class exist in a high-nasal-to-low-temporal gradient and ephrins of the B class are expressed higher dorsally than ventrally in the retina (Marcus et al. 1996). Such localisation of ligands in the eye also occurs in the zebrafish where three ephrins are differentially expressed in retinal ganglion cells prior to and during the projection of axons to the midbrain (Brennan et al. 1997).

Nakomoto et al. have provided direct evidence that Eph signalling is involved in retinotectal map formation by misexpressing ephrin-A2 in the developing chick tectum and showing that this leads to abnormal retinotectal axon growth (Nakamoto et al. 1996). Furthermore, a mouse carrying a mutation in the ephrin-A5 gene has abnormal projections of retinal ganglion cells to the tectum (Frisen et al. 1998). Although the gross aspects of the topographic map are normal in these mutants, axons that normally project to the caudal end of the tectum grow into the hindbrain region in the mutants. This result indicates that ephrin-A5 acts primarily as a block to axon growth. It is, however, not clear that all vertebrates will pattern the retinotectal projection in quite the same way. In the zebrafish, for example, there are three ligands present in the tectum, two of which have been shown to possess axon growth inhibitory properties (Brennan et al. 1997) and the zebrafish EphA4 homologue is absent from the eye (Xu et al. 1996).

The visual system is a complex of different axonal pathways and targets and it appears that some Eph receptors are involved in controlling the growth of specific subsets of these axonal projections. This has been demonstrated by the targeted mutation of the mouse EphA8 gene, which is normally expressed in a rostro-caudal gradient in the eye and in the superior colliculus. Mutant animals lack normal contralateral connections between the superior colliculi as well as connections from the superior colliculus to the spinal cord (Park et al. 1997).

The formation of topographic maps is not limited to the visual system and is a feature of other regions of the CNS such as the hippocampus and septum which are areas involved in learning and memory. It has been shown that an Eph signalling system may underlie the formation of topographic projections involving the septum and hippocampus (Gao et al. 1996; Zhang et al. 1996). The hippocampal neurons project to the lateral septum in a precise order and the hippocampus receives input from the medial septum. Ephrin-A2 is expressed in a dorsal-to-ventral gradient in the septum and, in culture, selectively allows growth of axons from appropriate regions of the hippocampus. The receptor EphA5, is expressed in a complementary lateral-to-medial gradient in the hippocampus.

Gene expression patterns suggest that Eph receptor signalling may be involved in establishing specific neuronal connections in the developing peripheral nervous system too. For example, the mouse receptor EphA3, and its rat and chick homologues are expressed in a subset of spinal motor neurons and a subset of axial muscles (Kilpatrick et al. 1996; Ohta et al. 1996). Furthermore, in the mouse, the ligand ephrin-A5 is expressed to a greater extent by head and neck muscles than by trunk and limb muscles, and muscle, cell lines derived from these different axial levels inhibit growth of dorsal root ganglion axons to different extents (Donoghue et al. 1996). These results suggest that specificity in connections of spinal motor and ganglionic neurons to the periphery may be based on Eph signalling.

A number of studies have now shown that Eph receptor/ephrin signalling leads to collapse of the neuronal growth cone, resulting in guidance of axon growth by inhibition. Using the stripe and growth cone collapse assays developed by Friedrich Bonhoeffer's laboratory it has been shown that the class A ephrins expressed in mouse, chick and zebrafish tectum cause growth cone collapse (Brennan et al. 1997; Drescher et al. 1995; Nakamoto et al. 1996). This is also the case for chick spinal motor neurons which express EphA4 and EphB2 and which collapse following interactions with ephrins of the A and B class in in vitro assays (Henkemeyer et al. 1994; Ohta et al. 1997; Wang and Anderson 1997).

Observations of growth cone behaviour in cultured CNS neurons following interactions with ephrins show that they collapse by withdrawal of filopodia (see, for example, Drescher et al. 1995). This has focused attention on the link between Eph receptor/class B ephrin signalling and the elements of the cytoskeleton. Evidence from in vitro assays and from targeted mutation studies indicates that signalling via Eph receptors or class B ephrins can lead to inhibition of growth cone advance (Henkemeyer et al. 1996). Recent work with cortical neurons in culture suggests that the responses of their growth cones to class A versus class B ephrins may be different in terms of the cytoskeletal components involved (Meima et al. 1997a,b). Interaction with ephrin-A5 leads to alterations in actin polymerisation in cortical growth cones whereas ephrin-B1 does not cause actin rearrangement but appears to affect microtubules in the growth cone.

8.3 Eph Signalling Is Involved in Controlling Neural Crest Cell Migration

In the trunk and in the head, neural crest cells take particular paths to reach the regions of the periphery in which they will settle and differentiate. In the trunk of the rat and chick embryo, for example, crest cells

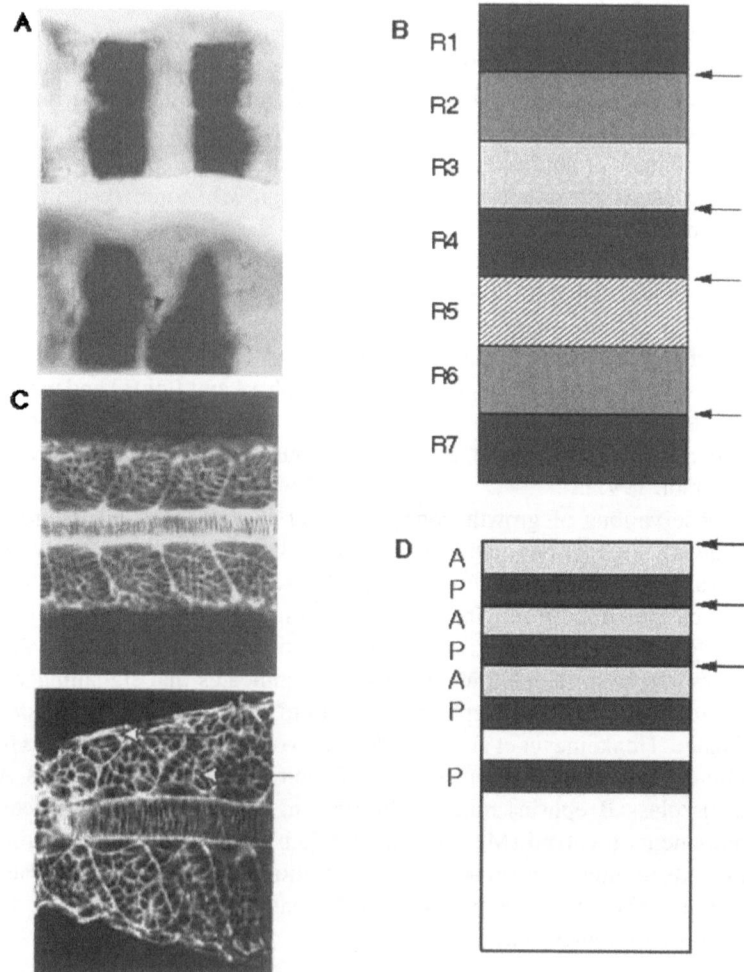

Fig. 1. Legend see p. 131

Fig. 1A–D. Eph/ephrin signalling and segmentation. **A** Interruption of EphA4 function following injection of RNA encoding a dominant-negative form of the receptor into the fertilised egg of the zebrafish leads to abnormal rhombomere boundary formation. In the *top panel* expression of the zebrafish receptor EphA4 is seen in the normal hindbrain in rhombomeres 3 and 5. *Below* is shown a hindbrain from an embryo into which synthetic RNA encoding an interfering form of the receptor has been injected. The rhombomere boundaries are abnormally formed and there remains a connection of EphA4-expressing cells between rhombomere 3 and 5 (*arrowhead*; after Xu et al. 1995). **B** Schematic representation of expression of ephrin-B2 (*dark shading* rhombomere1 (*R1*), *R4* and *R7*), EphA4 (*light shading R3*; *hatching R5*) and EphB4 (*medium shading R2, R6*; *hatching R5*). Ephrin-B2 activates both receptors and can account for segmental boundary formation at the sites of the *arrows*. **C** Somite boundary formation is abnormal in zebrafish embryos which have been injected with synthetic RNA encoding interfering forms of EphA4 or ephrin-B. Bodipy dye-stained zebrafish embryos showing the formation of the somites in normal embryos (*top panel*) and in an embryo injected with a synthetic RNA encoding a soluble form of ephrin-B2 (*bottom panel*). In the experimental embryo somite boundaries are abnormally formed (*arrows*; after Durbin et al. 1998). **D** Expression of ephrin-B2 (*dark shading*) in posterior halves (*P*) of the somite and in two bands in the unsegmented presomitic mesoderm, and EphA4 (*light shading*) in anterior halves (*A*) of somites and in a single band in the unsegmented presomitic mesoderm. Interactions at alternating interfaces between anterior and posterior half segments leads to boundary formation (*arrows*; based on Durbin et al. 1998)

are excluded from migrating through the caudal half of each somite. Two studies have now shown that this pattern of crest cell migration is due to an inhibition of crest cell movement through the caudal regions of the somite, a process that is mediated by Eph signalling. In the chick, EphB3 is expressed on crest cells and cells of the rostral half of the somite, whereas ephrin-B1, is expressed in the caudal half of the somite (Krull et al. 1997; Wang and Anderson 1997). In the rat, the receptor involved in this process is not clear, although the ephrin expressed in the caudal half of the somite is ephrin-B2, and is therefore a different class B ephrin to that performing the task in the chick.

A similar process of spatial exclusion underlies directed neural crest cell movement in the hindbrain where cells migrate from specific rhombomeres (see below) to specific branchial arches (Smith et al. 1997). This study concentrated on the migration of streams of neural crest from rhombomeres (r) 4, 5 and 6, which distribute crest cells to branchial arches 2, 3 and 4 respectively (Fig. 2). Expression of ephrin-B2 by crest cells from r4 prevents them from mixing with crest cells migrating from r5, which express EphA4 and EphB1, receptors that are both activated following binding to ephrin-B2.

Inhibition of migration through repulsive interactions is a theme underlying the roles of Eph receptors and ephrins in axon guidance and in neural crest cell migration. Time-lapse studies of neural crest cells have shown that contact between a crest cell and a cell carrying a repulsive signal results in the collapse of the crest cell filopodia in a similar manner to the axonal growth cone collapse seen in the retinotectal system (Jesuthasan 1996).

8.4 Function of Eph/Ephrin Signalling in the Formation of Boundaries in the Somites and Hindbrain

Segmentation is a basic process in embryogenesis of many invertebrate and all vertebrate embryos. In vertebrates the two regions of the body axis that are clearly segmented are the paraxial mesoderm, which gives rise to the somites – the precursors of the segmented vertebral column (Gossler and Hrabe de Angelis 1998) – and the hindbrain region of the neural plate. The hindbrain is divided up into regular units called rhombomeres, which are the basis for patterning of the neural epithelium and subsequent differentiation of neurons (Lumsden and Krumlauf 1996). In both regions of the embryo the segments develop clear boundaries at which the cells undergo distinctive behaviours involving cell shape changes. In the hindbrain, segmentation occurs within a defined region of the neural plate, whereas in the paraxial mesoderm segmentation is a dynamic process linked to the growth of the body axis at the posterior end. Eph receptors and ephrins are expressed in both hindbrain neural plate and paraxial mesoderm, and functional analysis of receptor signalling in zebrafish and in *Xenopus* indicates that such signalling is crucial for normal development of segment boundaries in both regions of the embryo (Fig. 2).

The vertebrate hindbrain consists of seven or eight rhombomeres, which become apparent during neurulation stages, once gastrulation is completed. Boundaries develop gradually and in a predictable sequence. The boundaries are evident because cells within the boundary zone have specific flattened shapes and they are organised in straight lines at right angles to the body axis (Heyman et al. 1994; Moens et al. 1998). These edges are barriers to cell movement and are extremes for expression of genes concerned with patterning the hindbrain (Lumsden and Krumlauf 1996). Eph receptors and ephrins are expressed in specific rhombomeres in such a way that receptors and ligands interact at the future boundaries. For example, in *Xenopus*, EphA4 is expressed in rhombomeres 3 and 5 (Nieto et al. 1992; Gale et al. 1996; Xu et al. 1995) and ephrin-B2 is expressed in rhombomeres 2, 4 and 6 (Smith et al. 1997). The fields of cells in these alternating rhombomeres interact only at the future boundaries between the rhombomeres. This interpretation is consistent with grafting experiments in the chick embryo in which it has been shown that interfaces between odd- and even-numbered rhombomeres are necessary for boundary formation to occur (Guthrie and Lumsden 1991). Interfering with Eph receptor/ephrin signalling following the injection of RNA encoding a dominant-negative form of EphA4 led to abnormal rhombomere boundary formation. In such experimental embryos rhombomeres have abnormal shapes and sizes with misplaced boundaries (Xu et al. 1995). The mechanism of EphA4 function in the developing hindbrain remains unclear although an analysis of the formation of boundaries in mouse and chick hindbrains shows that the process is gradual, suggesting that Eph signalling is important for the control of local cell movement (Irving et al. 1996).

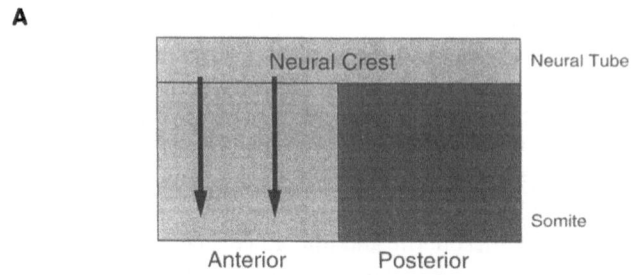

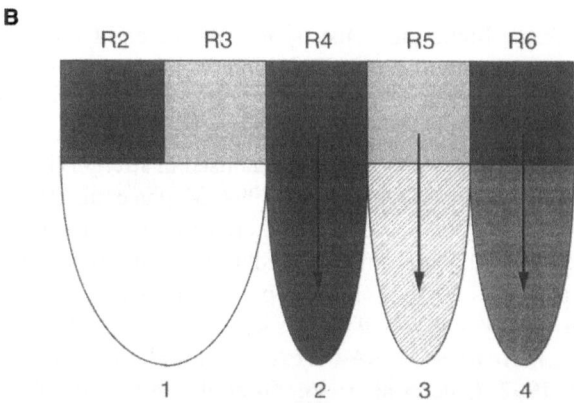

Fig. 2A,B. The role of Eph/ephrin signalling in controlling directed migration of the neural crest. **A** Summary drawing of the expression patterns of ephrins-B1 and B2 in the somitic tissue in the chick and rat (*dark shading*). In the chick the neural crest cells express EphB3 (*light shading*). Interactions between receptor and ligand cause repulsion of neural crest cell movement leading to migration of crest cells through the rostral half of the somite (based on Krull et al. 1997; Wang and Anderson 1997). **B** Migration of neural crest from the hindbrain region into the branchial arches is also controlled by Eph/ephrin signalling. In *Xenopus*, Ephrin-B2 (*dark shading*) is expressed in rhombomere 2 (*R2*), *R4* and *R6* and EphA4 (*light shading*) is expressed in *R3* and *R5*. Neural crest cells derived from these rhombomeres also express these ligands and receptors, and interactions between migrating neural crest cells along their migration pathways into the four branchial arches (*numbered*) ensures that neural crest cells from the appropriate rhombomere end up in the appropriate arch. The receptor EphB1 is also expressed in the crest of *R5* along with EphA4 (*hatching*), and EphB1 is expressed exclusively in *R6* neural crest (*medium shading*; based on Smith et al. 1997)

Recent results in the zebrafish embryo show that a similar process, based on the expression of alternating stripes of Eph receptors and ephrins, is important for normal somite segmentation to occur. Several Eph receptors and ephrins are expressed in the somitic mesoderm in a number of vertebrate species (Bergemann et al. 1995; Cooke et al. 1997; Flenniken et al. 1996; Gale et al. 1996; Scales et al. 1995). Using a dominant-negative strategy of injecting RNA encoding kinase-inactive receptors or soluble ephrins into the zebrafish embryo, it has been shown that Eph receptor/ephrin signalling is required for normal somite segmentation (Durbin et al. 1998). As in the hindbrain rhombomeres, boundaries in the paraxial mesoderm are misplaced or absent in experimental embryos.

There is an important difference between the spatial arrangement of Eph receptor/ephrin expression in the somites and in the hindbrain. In the latter, expression domains correspond to alternating rhombomeres, but in the somites, expression domains of Eph receptor and ephrin correspond to anterior and posterior halves of a single somite. Thus, in the paraxial mesoderm a somite border forms at alternate interfaces between receptor- and ligand-expressing cells. This is consistent with grafting experiments in the chick embryo in which it was shown that interfaces between anterior and posterior regions of the somite are required for a boundary to form (Stern and Keynes 1987). It is of considerable interest to know how these expression domains are controlled in the forming hindbrain and in the paraxial mesoderm. In the hindbrain it is known that the EphA4 expression in rhombomeres 3 and 5 is under the control of the transcriptional regulator Krox-20 (Theil et al. 1998). It is not yet clear how the dynamic expression of EphA4 and ephrin-B2 is controlled in the unsegmented presomitic mesoderm.

In addition to its restricted expression in the hindbrain, EphA4 is also expressed in distinct domains of the developing forebrain. Experiments in which a dominant-negative form of EphA4 was overexpressed demonstrated a role for EphA4 in the regionalisation of this tissue (Xu et al. 1996). The zebrafish EphA4 homologue is expressed from early neural plate stages in regions of the presumptive diencephalon that are fated not to become eye tissue. As development proceeds, the eye fields come to lie lateral to, and almost completely separate from, the diencephalon except for the location of the eye stalk. EphA4 expression persists in the ventral and dorsal diencephalic regions during these stages. In embryos injected with the dominant-negative EphA4 RNA, the forebrain regions fated to become ventral diencephalon become retina instead and large expanded eyes are formed. Again, the exact role of EphA4 in the regionalisation process is unclear, however, since extensive morphogenetic movements underlie the development of eye and ventral diencephalic tissue, an involvement of EphA4 in cell association or boundary formation is possible.

8.5 Principles of Eph/Ephrin Signalling

Eph receptor/ephrin signalling has been linked to a range of cellular responses including the control of cell movement, cell shape changes and cell growth. This raises questions as to how the specificity of response is achieved. For example, are different downstream signalling components expressed in different cell types? Also, how is the cytoskeleton stimulated differently to activate cell migration in one cell type but inhibit it in another? To begin to answer these questions it is necessary to understand the mechanisms of activation of receptors and class B ephrins, the structure of these molecules and the pathways which link the receptors and ephrin-B proteins to the intracellular signalling cascades.

Understanding the function of Eph signalling in the embryo demands a knowledge not only of the expression pattern but also the binding characteristics of the receptor and ligand pair involved. This is because there are variable affinities of binding within each Eph receptor and ephrin subclass (Brambilla et al. 1995; Brambilla et al. 1996; Gale et al. 1996; Lackmann et al. 1997; Monschau et al. 1997). This can be best illustrated with respect to a situation in the embryo where Eph signalling is known to be involved. One such case is the formation of the retinotectal projection where Eph signalling is required for the formation of the retinotopic map (Nakamoto et al. 1996). In the mouse, chick and zebrafish two class A ephrins, A2 and A5, are expressed in the tectum with graded distributions (Brennan et al. 1997; Cheng et al. 1995; Drescher et al. 1995; see above). In elucidating how the retinotectal map is created it is important to understand the binding characteristics of these two ligands with the receptors carried by the projection neurons, the retinal ganglion cells. One of these receptors is EphA3 and it has been shown recently, using alkaline phosphatase tagged proteins, that the dissociation constants are different for the interactions between EphA3 and ephrin-A5 and between EphA3 and ephrin-A2. Ephrin-A5 binds EphA3 with significantly greater efficacy than does ephrin-A2 (Monschau et al. 1997).

However, it is also important to compare the methods used to determine binding characteristics. For example, human EphA3 has similar affinity constants for binding to human ephrin-A3 and ephrin-A5 when the interaction is assessed with the ligands as Fc-fusion proteins, but the interaction is different when assessed by binding of monovalent ligands and soluble receptor. Under these conditions ephrin-A5 has a markedly lower dissociation rate from the soluble receptor than ephrin-A3, making it more likely that ephrin-A5 is the endogenous ligand (Lackmann et al. 1997).

A further important issue with regard to receptor activation is the observation by Stein et al. that the degree of oligomerisation of the ligand can affect the nature of the downstream response of the receptor (Stein et al. 1998). This goes some way towards an explanation for the specific responses of receptor-expressing cells which encounter fields of cells expressing graded distributions of ligand, such as in the retinotectal system described above.

The Eph receptors have a standard structure which is illustrated in Fig. 3. They have an uninterrupted catalytic domain intracellularly, and a cysteine-rich domain and two fibronectin type III repeats in the extracellular region. At the extracellular N-terminus there is a globular domain responsible for specificity of ligand binding (Labrador et al. 1997). This was shown by creating a series of soluble deletion and domain substitution mutants of EphB2 as alkaline phosphatase-tagged fusion proteins, and examining their binding to ephrin-B1. In domain deletion experiments, only EphB2 ectodomains containing the N-terminal globular domain bind to ephrin-B2. By switching the N-terminal globular region of EphB2 with the corresponding domain of the orphan receptor EphB5, it was shown that the EphB2 N-terminal globular domain was sufficient to confer ephrin-B1 specific binding. Also, the globular domain of EphA3 renders the EphB2 receptor competent to bind to the class A ligand, ephrin-A2 (Labrador et al. 1997). Furthermore, ephrin-B1-dependent transformation of NIH 3T3 cells was seen with chimaeric receptors in which the ectodomain of EphB2 was fused

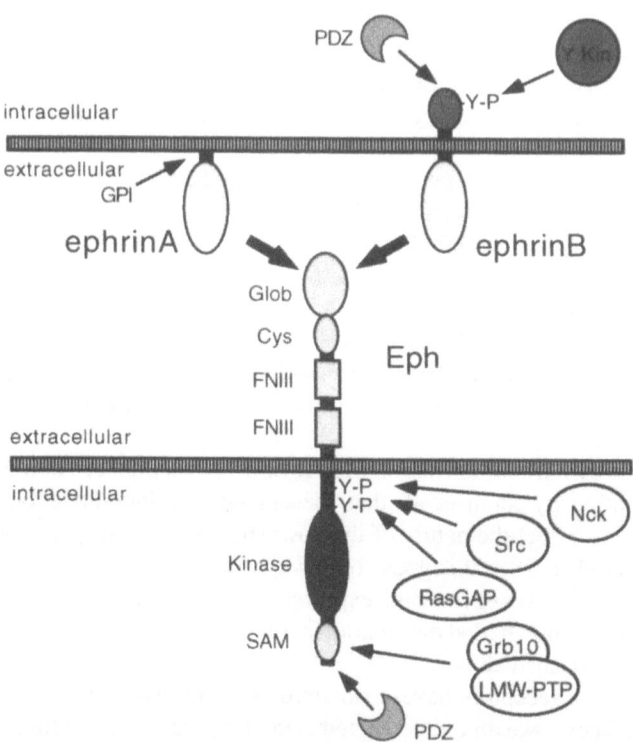

Fig. 3. Schematic representation of the domain structure, ephrin ligand interaction and signalling of Eph receptors. Both GPI-anchored ephrin-A and transmembrane ephrin-B ligands interact with the N-terminal globular domain (*Glob*) of Eph receptors. The globular domain is followed by a cysteine-rich region (*Cys*) and two fibronectin type III (*FNIII*) domains which contain a dimerisation motif. Nck, Src family kinases and RasGAP engage via two conserved tyrosine residues in the juxtamembrane region, Grb10 and LMW-PTP interact with a conserved tyrosine which is embedded in a SAM domain. PDZ domain proteins bind to C-terminal PDZ target sites in Eph receptors and ephrin-B molecules. Eph receptor contact induces tyrosine phosphorylation of the cytoplasmic domain of ephrin-B proteins via an as yet unknown tyrosine kinase (Y Kin; Based on Brückner and Klein 1998). (Colour versions of all of the figures in this paper have previously been accepted for publication [Holder and Klein 1999, with permission])

to the intracellular domain of the TrkB receptor tyrosine kinase. These results show conclusively that ligand-binding specificity resides in the N-terminal globular domain. Recently, the crystal structure of the N-terminal globular domain of EphB2 was solved (Himanen et al. 1998). The domain folds into a compact jellyroll β-sandwich composed of two antiparallel β-sheets and has structural similarities with the carbohydrate binding domain of lectins and influenza virus hemagglutinin. Structure-based mutagenesis identified an extended loop packed against the concave β-sandwich surface as important for ligand-binding and subclass specificity.

Adjacent to the N-terminal domain is an EGF-like, cysteine-rich region of unknown function and two fibronectin type III repeats. Such fibronectin type III repeats appear in ectodomains of numerous cell adhesion molecules, receptor tyrosine kinases and receptor tyrosine phosphatases and may be involved in dimerisation. In fact, incubation of cells with divalent complexes of the EGF-like region plus the two fibronectin type III repeats of EphA3 caused ligand-independent EphA3 receptor transphosphorylation suggesting the presence of a dimerisation motif (Lackmann et al. 1998). It was suggested that Eph receptor activation occurs by a two-step mechanism, with distinct ligand binding via the N-terminal globular domain followed by receptor-receptor oligomerisation via the more C-terminal dimerisation domain (Lackmann et al. 1998).

The C-terminal intracellular region of the protein includes the kinase domain. A highly conserved motif containing two tyrosine residues is found in the juxtamembrane intracellular region of all Eph receptors (Ellis et al. 1996; Holland et al. 1997). These tyrosine residues are also major in vitro autophosphorylation sites for EphA4 (Ellis et al. 1996) and EphB2 (Holland et al. 1997), and are likely to be important for intracellular signalling. It has been shown that a number of SH2 domain cytoplasmic proteins bind to the juxtamembrane region of the receptor when it is activated. These include the Src-like tyrosine kinases p59fyn and p60src which bind to this region in EphA4 (Ellis et al. 1996) and EphB2, respectively (Zisch et al. 1998). The Ras GTPase activating protein (RasGAP) binds through its SH2 domain to tyrosine phosphorylated EphB2, as does a 62–64 kDa protein p62dok and the SH2/SH3 domain adaptor protein Nck. It is likely that the RasGAP, p62dok and Nck proteins form a complex bound to the juxtamembrane region of

EphB2 and they potentially link signalling to control of cytoskeletal dynamics (Holland et al. 1997; Brückner and Klein 1998).

C-terminal to the kinase domain, a conserved region of 60–70 amino acids is present in all Eph receptors and was identified as a sterile alpha motif (SAM) domain (Schultz et al. 1997). An invariant tyrosine located within the SAM domain of EphB1 is required for binding of the Grb10 adaptor protein (Stein et al. 1996). It is of interest that Grb10 shares homology with a *Caenorhabditis elegans* gene product thought to be involved in neural cell migration (Ooi et al. 1995). The tyrosine within the SAM domain of EphB1 is also required for binding of a low molecular weight phosphotyrosine phosphatase (LMW-PTP) whose recruitment correlates with functional responses such as endothelial capillary-like assembly and cell attachment after stimulation with higher order ephrin clusters (Stein et al. 1998).

Finally, a PDZ-binding motif which interacts with PDZ domain proteins is present at the C-terminal tail of Eph receptors (PDZ for postsynaptic density protein, discs large, zona occludens; Sheng 1996). In line with their known interactions with synaptic membrane proteins, PDZ domain proteins were found to cluster and co-localize with Eph receptors at synapses of cultured hippocampal neurons (Torres et al. 1998). Some PDZ domain proteins become tyrosine phosphorylated when complexed with Eph receptors (Torres et al. 1998) and an intact Eph kinase domain appears to be required for the interaction (Hock et al. 1998). Interestingly, a functional PDZ-binding motif is also present at the C-terminus of transmembrane ephrin-B proteins (Torres et al. 1998). PDZ domain proteins may therefore be important mediators of ephrin clustering and/or signalling.

8.6 Conclusions

Eph receptors and ephrins are dynamically expressed during development of a range of vertebrate species and have been isolated in *C. elegans* (George et al. 1998). In this review I have outlined the role that Eph/ephrin signalling plays in axon guidance, neural crest migration and segmentation to illustrate the importance of these signalling proteins to fundamental morphogenetic mechanisms in the embryo. The examples are principally chosen from development of the nervous system but it is also important to be aware that Eph/ephrin signalling is centrally involved in other developmental processes. These include gastrulation, where there is growing evidence for a role for Eph/ephrin signalling in the primary movement of cells during convergence and extension of the vertebrate embryo (Winning et al. 1996; Jones et al. 1998). It is also clear that Eph/ephrin signalling is involved in the formation of blood vessels within the embryo (Wang et al. 1998), as ephrin-B2 loss-of-function mutants, produced by targeted disruption of the mouse ephrin-B2 gene, have an abnormal embryonic vasculature. Examination of Eph/ephrin function with regard to these diverse processes indicates that signalling through these receptor tyrosine kinases and the class B ephrins controls cellular morphology. In the case of the growth cone and the neural crest cell, guidance is achieved by repulsive interactions between receptor-expressing and ligand-expressing cells which occupy reciprocal spatial domains in the tissues of the embryo. Questions remain concerning the intracellular pathways linking Eph receptor/ephrin signalling to the cytoskeleton. Our understanding of the specificity of cellular responses to this large group of receptors will be resolved as the intracellular pathways are defined and the structural features of the receptor and ligand proteins are understood. Little is known about the upstream regulation of the Eph receptor and ephrin genes. The dynamic nature of their expression and the changes in expression of different members of the class A and class B families between species pose interesting questions in this regard.

References

Bartley TD, Hunt RW, Welcher AA, Boyle WJ, Parker VP, Lindberg RA, Lu HS, Colombero AM, Elliot RA, Guthrie BA, Holst PL, Skrine JD, Toso RJ, Zhang M, Fernandez E, Trail G, Hunter T, Fox GM (1994) B61 is a ligand for the ECK receptor protein-tyrosine kinase. Nature 368:558–560

Bergemann A, Hwai-Jong C, Brambilla R, Klein R, Flanagan J (1995) Elf-2, a new member of the Eph ligand family, is segmentally expressed in mouse embryos in the region of the hindbrain and newly forming somites. Mol Cell Biol 15:4921–4929

Brambilla R, Schnapp A, Casagranda F, Labrador J, Bergeman A, Flanagan J, Pasquale E, Klein R (1995) Membrane bound LERK2 ligand can signal through three different Eph-related receptor tyrosine kinases. EMBO J 14:3116–3126

Brambilla R, Bruckner K, Orioli D, Bergemann A, Flanagan J, Klein R (1996) Similarities and differences in the way transmembrane-type ligands interact with the Elk subclass of Eph receptors. Mol Cell Neurosci 8:199–209

Brandli A, Kirschner M (1995) Molecular cloning of tyrosine kinases in the early *Xenopus* embryo: identification of Eck-related genes expressed in cranial neural crest cells of the second (hyoid) arch. Dev Dyn 203:119–140

Brennan C, Monshau B, Lindberg R, Guthrie B, Drescher U, Bonhoeffer F, Holder N (1997) Two Eph receptor tyrosine kinase ligands control axon growth and may be involved in the creation of the retinotectal map in zebrafish. Development 124:655–664

Brückner K, Klein R (1998) Signalling by Eph receptors and their ephrin ligands. Curr Opin Neurobiol 8:375–382

Bruckner K, Pasquale E, Klein R (1997) Tyrosine phosphorylation of transmembrane ligands for Eph receptors. Science 275:1640–1643

Cheng H-J, Flanagan J (1994) Identification and cloning of ELF-1, a developmentally expressed ligand for the Mek4 and Sek 1 receptor tyrosine kinases. Cell 79:157–168

Cheng H-J, Nakamoto M, Bergemann A, Flanagan, J (1995) Complementary gradients in expression and binding of Elf-1 and Mek4 in development of the topographic retinotectal projection map. Cell 82: 371–381

Cooke J, Xu Q, Wilson S, Holder N (1997) Characterisation of five novel zebrafish Eph-related receptor tyrosine kinases suggests roles in neural patterning. Dev Genes Evol 206:515–531

Davis S, Gale N, Aldrich T, Maisonpierre P, Lhotak V, Pawson T, Goldfarb M, Yancopoulos G (1994) Ligands for EPH-related receptor tyrosine kinases that require membrane attachment or clustering for activity. Science 266:816–819

Donoghue M, Lewis R, Merlie J, Sanes J (1996) The eph kinase ligand AL-1 is expressed by rostral muscles and inhibits outgrowth from caudal muscles. Mol Cell Neurosci 8:185–198

Drescher U, Kremoser C, Handwerker C, Loschinger J, Noda M, Bonhoeffer F (1995) In vitro guidance of retinal ganglion cell axons by RAGS, a 25 kDa tectal protein related to the ligands for Eph receptor tyrosine kinases. Cell 82:359–370

Durbin L, Brennan C, Shiomi K, Cooke J, Barrios A, Shanmugalingam S, Guthrie B, Lindberg R, Holder N (1998) Eph signalling is required for segmentation and differentiation of somites. Genes Dev 12:3096–3109

Ellis C, Kasmi F, Ganju P, Walls E, Panayotou G, Reith A (1996) A juxtamembrane autophosphorylation site in the Eph family receptor tyrosine kinase,

Sek, mediates high affinity interaction with p59fyn. Oncogene 12:1727–1736

Eph Nomenclature Committee (1997) Unified nomenclature for Eph family receptors and their ligands. Cell 90:403

Flenniken A, Gale N, Yancopoulos G, Wilkinson D (1996) Distinct and overlapping expression patterns of ligands for Eph related receptor tyrosine kinases during mouse development. Dev Biol 179:382–401

Frisen J, Yates P, McLaughlin T, Friedman C, O'Leary D, Barbacid M (1998) Ephrin-A5 (AL-1/RAGS) is essential for proper retinal axon guidance and topographic mapping in the mammalian visual system. Neuron 20:235–243

Gale N, Holland S, Valenzuela D, Flenniken A, Pan L, Ryan T, Henkemeyer M, Strebhardt K, Hirai H, Wilkinson D, Pawson T, Davis S, Yancopoulos G (1996) Eph receptors and ligands comprise two major specificity subclasses and are reciprocally compartmentalized during embryogenesis. Neuron 17:9–19

Gao P, Zhang J, Yokoyama M, Racey B, Dreyfus C, Black I, Zhou R (1996) Regulation of topographic projection in the brain: elf-1 in the hippocamposeptal system. Proc Natl Acad Sci USA 93:11161–11166

George S, Simokat K, Hardin J, Chisholm A (1998) The VAB-1 Eph receptor tyrosine kinase functions in neural and epithelial morphogenesis in *C. elegans*. Cell 92:633–643

Gossler A, Hrabe de Angelis M (1998) Somitogenesis. Current Topics in Dev Biol 38:225–287

Guthrie S, Lumsden A (1991) Formation and regeneration of rhombomere boundaries in the developing chick hindbrain. Development 112:221–229

Henkemeyer M, Marengere L, McGlade J, Olivier J, Conlon R, Holmyard D, Letwin K, Pawson T (1994) Immunolocalisation of the Nuk receptor tyrosine kinase suggests roles in segmental patterning of the brain and axonogenesis. Oncogene 9:1001–1014

Henkemeyer M, Orioli D, Henderson J, Saxton T, Roder J, Pawson T, Klein R (1996) Nuk controls pathfinding of commissural axons in the mammalian central nervous system. Cell 86:35–46

Heyman I, Kent A, Lumsden A (1994) Cellular morphology and extracellular space at rhombomere boundaries in the chick embryo hindbrain. Dev Dyn 198:241–253

Himanen J-P, Henkemeyer M, Nikolov D (1998) Crystal structure of the ligand-binding domain of the receptor tyrosine kinase EphB2. Nature 396:486–491

Hirai H, Maru Y, Hagiwara K, Nishida J, Takaku F (1987) A novel putative tyrosine kinase receptor encoded by the eph gene. Science 238:1717–1720

Hock B, Böhme B, Karn T, Yamamoto T, Kaibuchi K, Holtrich U, Holland S, Pawson T, Rübsamen-Waigmann H, Strebhardt K (1998) PDZ-domain-mediated interaction of the Eph-related receptor tyrosine kinase EphB3 and the ras-binding protein AF6 depends on the kinase activity of the receptor. Proc Natl Acad Sci USA 95:9779–9784

Holash J, Pasquale E (1995) Polarized expression of the receptor protein tyrosine kinase Cek5 in the developing avian visual system. Dev Biol 172:683–693

Holder N and Klein R (1999) Eph receptors and ephrins: effectors of morphogenesis. Development 126: 2033–2044

Holland S, Gale N, Mbamalu G, Yancopoulos G, Henkemeyer M, Pawson T (1996) Bidirectional signalling through the Eph-family receptor Nuk and its transmembrane ligands. Nature 383:722–725

Holland S, Gale N, Gish G, Roth R, Songyang Z, Cantley L, Henkemeyer M, Yancopoulos G, Pawson T (1997) Juxtamembrane tyrosine residues couple the Eph family receptor EphB2/Nuk to specific SH2 domain proteins in neuronal cells. EMBO J 16:3877–3888

Irving C, Nieto A, DasGupta R, Charnay P, Wilkinson D (1996) Progressive spatial restriction of Sek-1 and krox-20 gene expression during hindbrain segmentation. Dev Biol 173:26–38

Jesuthasan S (1996) Contact inhibition/collapse and pathfinding of neural crest cells in the zebrafish trunk. Development 122:381–389

Jones T, Chong L, Kim J, Xu R, Kung H, Daar I (1998) Loss of cell adhesion in *Xenopus* laevis embryos mediated by the cytoplasmic domain of XLerk, an erythropoietin-producing hepatocellular ligand. Proc Natl Acad Sci USA 95:576–581

Kenny D, Bronner-Fraser M, Marcelle C (1995) The receptor tyrosine kinase QEK5 mRNA is expressed in a gradient within the neural retina and the tectum. Dev Biol 172:708–716

Kilpatrick T, Brown A, Lai C, Gassman M, Goulding M, Lemke G (1996) Expression of the Tyro4/Mek4/Cek4 gene specifically marks a subset of embryonic motor neurons and their muscle targets. Mol Cell Neurosci 7:62–74

Krull C, Lansford R, Gale N, Collazo A, Marcelle C, Yancopoulos G, Fraser S, Bronner-Fraser M (1997) Interactions of Eph-related receptors and ligands confer rostrocaudal pattern to trunk neural crest migration. Curr Biol 7:571–580

Labrador J, Brambilla R, Klein R (1997) The N-terminal globular domain of Eph receptors is sufficient for ligand binding and receptor signalling. EMBO J 16:3889–3897

Lackmann M, Mann R, Kravets L, Smith F, Bucci T, Maxwell K, Howlett G, Olsson J, Vanden Bos T, Cerretti D, Boyd A (1997) Ligand for Eph-related kinase (LERK) 7 is the preferred high affinity ligand for the Hek receptor. J Biol Chem 272:16521–16530

Lackmann M, Oates A, Dottori M, Smith F, Do C, Power M, Kravets L, Boyd A (1998) Distinct subdomains of the EphA3 receptor mediate ligand binding and receptor dimerization. J Biol Chem 273:20228–20237

Lumsden A, Krumlauf R (1996) Patterning the vertebrate neuraxis. Science 274:1109–1115

Marcus R, Gale N, Morrison M, Mason C, Yancopoulos G (1996) Eph family receptors and their ligands distribute in opposing gradients in the developing mouse retina. Dev Biol 180:786–789

Meima L, Kljavin I, Moran P, Shih A, Winslow J, Carras I (1997a) AL-1 induced growth cone collapse of rat cortical neurons is correlated with REK-7 expression and rearrangement of the actin cytoskeleton. Eur J Neurosci 9:177–188

Meima L, Moran P, Mathews W, Carras, I (1997b) Lerk2 (ephrin-B1) is a collapsing factor of a subset of cortical growth cones and acts by a mechanism different from AL-1 (ephrin-A5). Mol Cell Neurosci 9:314–328

Moens C, Cordes S, Giorgianni M, Barsh G, Kimmel C (1998) Equivalence in the genetic control of hindbrain segmentation in fish and mouse. Development 125:381–391

Monschau B, Kremoser C, Ohta K, Tanaka H, Kaneko T, Yamada T, Handwerker C, Hornberger M, Loschinger J, Pasquale E, Siever D, Verderame M, Muller B, Bonhoeffer F, Drescher U (1997) Shared and distinct functions of RAGS and ELF-1 in guiding retinal axons. EMBO J 16:1258–1267

Nakamoto M, Cheng H-J, Friedman G, McLaughlin T, Hansen M, Yoon C, O'Leary D, Flanagan J (1996) Topographically specific effects of ELF-1 on retinal axon guidance in vitro and retinal axon mapping in vivo. Cell 86:755–766

Nieto MA, Gilardi-Hebenstreit P, Charnay P, Wilkinson DG (1992) A receptor protein tyrosine kinase implicated in the segmental patterning of the hindbrain and mesoderm. Development 116:1137–1150

Ohta K, Nakamura M, Hirokawa K, Tanaka S, Iwana A, Suda T, Ando M, Tanaka H (1996) The receptor tyrosine kinase, Cek 8, is transiently expressed on subtypes of motor neurons in the spinal cord during development. Mech Dev 54:59–69

Ohta K, Iwamasa H, Drescher U, Terasaki H, Tanaka H (1997) The inhibitory effect on neurite outgrowth of motoneurons exerted by the ligands ELF-1 and RAGS. Mech Dev 64:127–135

Ooi J, Yajnik V, Immanuel D, Gordon M, Moskow J, Buchberg A, Margolis B (1995) The cloning of Grb10 reveals a new family of Sh2 domain proteins. Oncogene 10:1621–1630

Orioli D, Henkemeyer M, Lemke G, Klein R, Pawson T (1996) Sek4 and Nuk receptors cooperate in guidance of commissural axons and in palate formation. EMBO J 15:6035–6049.

Orioli D, Klein R (1997) The Eph receptor family: axonal guidance by contact repulsion. Trends Genet 13:354–359

Pandey A, Lindberg R, Dixit V (1995) Receptor orphans find a family. Curr Biol 5:986–989

Park S, Frisen J, Barbacid M (1997) Aberrant axonal projections in mice lacking EphA8 (Eek) tyrosine kinase receptors. EMBO J 16:3106–3114

Pasquale E (1997) The Eph family of receptors. Curr Opin Cell Biol 9:608–615

Pasquale E, Connor R, Rochcoll D, Schburch H, Risau W (1994) Cek5, a tyrosine kinase of the Eph subclass, is activated during neural retina differentiation. Dev Biol 163:491–502

Pasquale EB, Deerinck TJ, Singer SJ, Ellisman MH (1992) Cek5, a membrane receptor-type tyrosine kinase, is in neurons of the embryonic and postnatal avian brain. J Neurosci 12:3956–3967

Scales J, Winning R, Renaud C, Shea L, Sargent T (1995) Novel members of the eph receptor kinase subfamily expressed during *Xenopus* development. Oncogene 11:1745–1752

Schultz J, Ponting CP, Hofmann K, Bork P (1997) SAM as a protein interaction domain involved in developmental regulation. Protein Sci 6:249–253

Sheng M (1996) PDZs and receptor/channel clustering: rounding up the latest suspects. Neuron 17:575–578

Smith A, Robinson V. Patel K, Wilkinson D (1997) The EphA4 and EphB1 receptor tyrosine kinases and ephrin-B2 ligand regulate targeted migration of branchial neural crest cells. Curr Biol 7:561–570

Stein E, Cerretti P, Daniel T (1996) Ligand activation of ELK receptor tyrosine kinase promotes its association with Grb10 and Grb2 in vascular endothelial cells. J Biol Chem 271:23588–23593

Stein E, Lane A, Cerretti D, Schoecklmann H, Schroff A, Van Etten R, Daniel T (1998) Eph receptors discriminate specific ligand oligomers to determine alternative signalling complexes, attachment, and assembly responses. Genes Dev 12:667–678

Stern C, Keynes R (1987) Interactions between somite cells: the formation and maintenance of segment boundaries in the chick embryo. Development 99:261–272

Theil T, Frain M, Gilardi-Hebenstreit P, Flenniken A, Charnay P, Wilkinson P (1998) Segmental expression of the EphA4 (Sek-1) receptor tyrosine kinase in the hindbrain is under direct transcriptional control of Krox-20. Development 125:443–452

Torres R, Firestein BL, Dong H, Staudinger J, Olson EN, Huganir RL, Bredt DS, Gale NW, Yancopoulos GD (1998) PDZ proteins bind, cluster and synaptically co-localize with Eph receptors and their ephrin ligands. Neuron 21:1453–1463

Tuzi N, Gullick W (1994) Eph, the largest known family of putative growth factor receptors. Br J Cancer 69:417–421

Van der Geer P, Hunter T, Lindberg R (1994) Receptor protein-tyrosine kinases and their signal transduction pathways. Annu Rev Cell Biol 10:251–337

Wang H, Anderson D (1997) Eph family transmembrane ligands can mediate repulsive guidance of trunk neural crest migration and motor axon outgrowth. Neuron 18:383–396

Wang H, Anderson D (1998) Molecular distinction and angiogenic interaction between embryonic arteries and veins revealed by ephrin-B2 and its receptor Eph-A4. Cell 93:741–753

Weinstein D, Rahman S, Ruiz J and Hemmati-Brivanlou A (1996) Embryonic expression of eph signalling factors in *Xenopus*. Mech Dev 57:133–144

Winning RS, Scales J, Sargent T (1996) Disruption of cell adhesion in *Xenopus* embryos by Pagliaccio, an Eph-class receptor tyrosine kinase. Dev Biol 179:309–319

Winslow J, Moran P, Valverde J, Shih A, Yuan J, Wong S, Tsai S, Goddard A. Henzel W, Hefti F, Beck K, Caras I (1995) Cloning of AL-1, a ligand for an Eph-related tyrosine kinase receptor involved in axon bundle formation. Neuron 14:973–981

Xu Q, Alldus G, Holder N, Wilkinson DG (1995) Expression of truncated Sek-1 receptor tyrosine kinase disrupts the segmental restriction of gene expression in the *Xenopus* and zebrafish hindbrain. Development 121:4005–4016

Xu Q, Alldus G, Macdonald R, Wilkinson D, Holder N (1996) Function of the Eph-related receptor tyrosine kinase gene rtk1 is required for regional specification in the zebrafish forebrain. Nature 381:319–322

Zhang J-H, Cerretti D, Yu T, Flanagan J, Zhou R (1996) Detection of ligands in regions anatomically connected to neurons expressing the Eph receptor Bsk: potential roles in neuron-target interaction. J Neurosci 16:7182–7192

Zisch A, Kalo MS, Chong LD, Pasquale EB (1998) Complex formation between EphB2 and Src requires phosphorylation of tyrosine 611 in the EphB2 juxtamembrane region. Oncogene 20:2657–2670

Obituary

Nigel Holder (1953–1998)

Nigel Holder died on 11 December 1998 at the tragically young age of 45. Nigel was a much-loved and respected scientist whose influence was diverse and far reaching. He led an internationally respected research group, established the highly successful Developmental Biology Research Centre at Kings College London and more recently, headed the Department of Anatomy and Developmental Biology at University College London, one of the UK's largest and most successful life science departments.

Nigel's research was directed at understanding the mechanisms that underlie the patterning of cells and tissues in the vertebrate embryo. In recent years, he had become interested in the mechanisms responsible for patterning of hindbrain segments and of somites in the developing zebrafish embryo. Much of his research focused on the role of the Eph family of molecules, which appear to regulate the behaviour of cells at the boundaries between segments.

There are many ways to be a successful scientist: Nigel chose the best way – to have endless enthusiasm and to take real pleasure from doing science. He encouraged cooperation and collaboration whenever possible and indeed in the last 10 years, all 30 of his research papers were collaborative efforts. Nigel was great fun to be around and his scientific colleagues frequently became close personal friends. His enthusiasm, humour, inventiveness and energy were infectious and made him the most treasured of colleagues. Nigel had that rare gift of genuinely caring about other people, their worries, their successes and their failures. Nigel had the ability to make people feel at ease and this made him a great communicator and leader. He used his intelligence, power and influence wisely and universally gained the trust of those around him.

In 1992, Nigel was diagnosed with a rare form of vasculitis, which at times was terribly debilitating and painful. He continued to work with undiminished enthusiasm and it is all the more remarkable that many of his major scientific achievements were attained despite the disease. He coped with his illness with bravery, dignity and, as ever, with humour. It is a tragic irony that in the last year, Nigel's health had improved considerably and he appeared to be in better shape than at any time since contracting the disease.

A Commemoration Meeting was held on 18 January 1999, the date that Nigel was due to give his inaugural lecture as Professor and Head of Anatomy and Developmental Biology at UCL. This very touching and poignant meeting combined science with personal recollection and was a fitting tribute to an influential and greatly admired scientist.

Stephen Wilson
Department of Anatomy and Developmental Biology,
University College London

9 Eph Receptor Tyrosine Kinases and Their Ligands in Development

U. Drescher

9.1 Introduction

A current focus of interest in developmental neurobiology is the identification and characterization of mechanisms and molecules contributing to the formation of neuronal connections. A number of gene families playing a role in this process have recently been identified. They can be classified tentatively according to the basic mechanisms by which axonal pathfinding to, and guidance in, the target area are thought to occur: chemoattraction by Netrins, chemorepulsion by Netrins and secreted Semaphorins; contact-mediated attraction by Ig domain-containing cell adhesion molecules (Ig-CAMs), Cadherins, and extracellular matrix proteins (ECM); and finally, contact-mediated repulsion by transmembrane Semaphorins, ECM proteins, and Ephrins (Tessier-Lavigne and Goodman 1996; Müller 1999).

Members of the Eph family – the Eph receptor tyrosine kinases and their ligands the Ephrins – are predominantly and widely expressed in the developing and adult nervous system (Zhou 1998). In early development, Ephrins and the corresponding Eph receptors are often found expressed complementarily, subdividing the embryo into broad structural domains (Gale et al. 1996). As the Eph family appears to exert its function by a repellent mechanism, the complementary expression of ligands and receptors indicates a patterning function of this family; for example, the growth of axons, or the movement of cells, expressing a certain receptor is restricted to regions of the embryo devoid of the corresponding ligand class.

In recent years, the Eph family has been shown to be involved in a number of processes during neural development, the development of topographic projections, and the segmental restriction of neural crest cells, motor axons, and hindbrain segments, to name just a few. Processes outside the nervous system in which this gene family exerts its function include angiogenesis.

9.2 Receptor–Ligand Interactions in the Eph Family

At present, 14 different receptors and 8 ligands are recognized for the Eph family, subdivided into 2 classes on the basis of both sequence homologies and binding specificities: the EphA subclass containing glycosyl-phosphatidylinositol (GPI) – anchored EphrinA ligands interacting with EphA receptors and the EphB subclass with transmembrane-anchored EphrinB ligands interacting with a complementary set of EphB receptors. This subdivision is followed relatively strictly, and few exceptions to this classification, such as the EphA4 receptor which is able to bind to both classes of Ephrin ligands, have been reported.

Within each of the two subclasses, the interaction between receptors and ligands appears to be highly promiscuous, e.g., each receptor is able to bind virtually all ligands of the corresponding subclass and vice versa. This is indicated, for example, by the fact that soluble receptor or ligand affinity probes (RAP, LAP; Flanagan and Leder 1990) often pick up the entire set of ligands or receptors during staining of sections. However, on detailed examination, there is partially considerable variation in the binding affinities between receptors and ligands (e.g., Gale et al. 1996; Monschau et al. 1997); thus the high degree of promiscuity in the interaction of ligands and receptors might be limited in vivo.

Binding affinities have often been determined by measuring the interaction between membrane-bound receptors and (artificially generated) soluble ligands and vice versa. In vivo, Eph ligands are membrane-bound (being a necessary prerequisite for their function), thus the binding affinities measured possibly do not reflect the real specificity of interaction. In support of this notion, in some cases high affinity receptor–ligand interactions did not correlate with the induction of a physiological response (e.g., Brambilla et al. 1996).

The level of clustering of Eph ligands, i.e., dimeric vs multimeric forms, also plays an important role with respect to the mode of intracellular signaling of the corresponding Eph receptors (Stein et al. 1998). Although dimeric EphrinB2 ligand induced tyrosine phosphorylation of the corresponding receptor, it did not trigger a physiological response. In contrast, the tetrameric form of the same ligand led to both receptor tyrosine phosphorylation and induction of a physiological response in the receptor-bearing cells. These differences in activity were attributed to the binding and activation of a low-molecular weight protein tyrosine phosphatase (LMW-PTP) to the activated receptor complex which is induced by the tetrameric, but not the dimeric ligand.

In sum, it appears that the functional characterization of this family is still impeded by our poor understanding of the mode and specificity of Eph receptor–ligand interactions.

9.3 The Retinotectal Projection

The retinotectal projection is a well-characterized system (Mey and Thanos 1992; Holt and Harris 1993) for studying the function of the Eph family, serving as a model for topographic projections which are numerous in the central and peripheral nervous system. In such projections there is a faithful transfer of spatially organized information from one area of the brain to another, so that cells neighboring in the projecting area are connected to cells neighboring in the target area. In the retinotectal projection, temporal retina is connected to rostral tectum and nasal retina to caudal tectum; similarly, dorsal and ventral retina are connected to ventral and dorsal tectum, respectively. How such precise connections are formed has been a longstanding puzzle. The presently prevailing hypothesis, formulated by Sperry decades ago, is the chemoaffinity hypothesis (Sperry 1963). Sperry proposed that molecules expressed in gradients in the tectum would provide positional information along which ingrowing retinal axons, endowed with receptors expressed in gradients, might find their correct retinotopic position.

Both EphrinA2 and EphrinA5 are expressed in overlapping high-caudal-to-low-rostral gradients in the tectum (Monschau et al. 1997), and a corresponding receptor, EphA3, is found in a high-temporal-to-low-nasal expression in the retina (Cheng et al. 1995). Their complementary patterns make these molecules likely candidates for an involvement in the topographic mapping of retinal axons as further supported by comparatively high binding affinities between EphA3 and EphrinA2/EphrinA5 (Monschau et al. 1997). Besides these three molecules there are at least nine other Eph family members expressed during development of the retinotectal projection, often in dynamic spatially and temporally restricted patterns (for review see Drescher et al. 1997; Pasquale 1997; Flanagan and Vanderhaeghen 1998; O'Leary and Wilkinson 1999).

In vitro, EphrinA5 repels both nasal and temporal axons, the latter being more sensitive than the former (Monschau et al. 1997). EphrinA2 exhibits a repellent activity exclusively for temporal axons, with no effect on nasal axons (Nakamoto et al. 1996; Monschau et al. 1997). In vivo, temporal but not nasal axons avoid patches of ectopically expressed EphrinA2 (Nakamoto et al. 1996). A loss-of-function analysis showed that EphrinA5 is essential for the correct topographic mapping of retinal axons in the mammalian visual system (Frisen et al. 1998). Here, temporal axons formed termination zones not only in the rostral, but also in the topographically inappropriate caudal superior colliculus, a structure homologous to the chick optic tectum. Additionally, a transient overshooting of retinal axons into the inferior colliculi in these mutant mice points to a function of EphrinA5 in limiting the growth of retinal axons to the superior colliculi.

Interestingly, EphrinA2 and EphrinA5 expression is seen not only in the tectum, but also in the retina (Marcus et al. 1996; Brennan et al. 1997; Connor et al. 1998), where they are expressed in a high-nasal-to-low-temporal pattern on retinal ganglion cell axons, thus colocalizing with EphA receptors (Fig. 1; Hornberger et al. 1999). In the stripe assay,

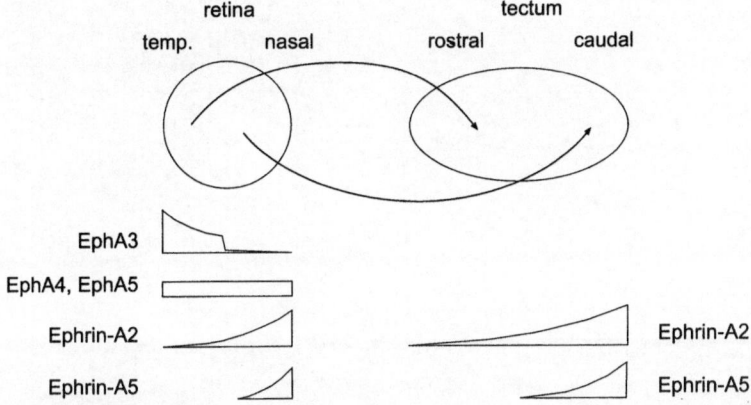

Fig. 1. Projection scheme of retinal axons and a summary of the expression pattern of EphA family members in the retina and the tectum. The EphA4 and EphA5 receptors are uniformly expressed in the retina, in contrast to the EphA3 receptor, which is expressed in the temporal (*temp.*) retina in a gradient with no or little expression in the nasal retina. Ephrin-A2 and Ephrin-A5 are expressed in the retina in a high-nasal-to-low-temporal gradient and in the tectum in a high-caudal-to-low-rostral gradient. The projection of temporal axons onto the rostral tectum and the projection of nasal axons onto the caudal tectum are indicated

only temporal axons are normally sensitive for repellent axon guidance cues from the caudal tectum (see above). However, retrovirally driven overexpression of either EphrinA2 or EphrinA5 on temporal retinal axons abolished the sensitivity for the repulsive activity of the caudal tectum (Fig. 2), whereas treatment of retinal axons with phosphatidylinositol-specific phospholipase C (PI-PLC) both removed the GPI-anchored EphrinA ligands and induced a striped outgrowth of formerly insensitive nasal axons.

In an attempt to find a possible biochemical correlate for this altered guidance behavior, the tyrosine phosphorylation pattern of EphA receptors on retinal ganglion cell axons was investigated. For this purpose the EphA4 receptor was chosen, which apparently is uniformly expressed in the retina and on retinal ganglion cell axons. In normal and control infected retinas, the EphA4 receptor is tyrosine phosphorylated in the nasal area and only faintly in the temporal area. However, in retinas

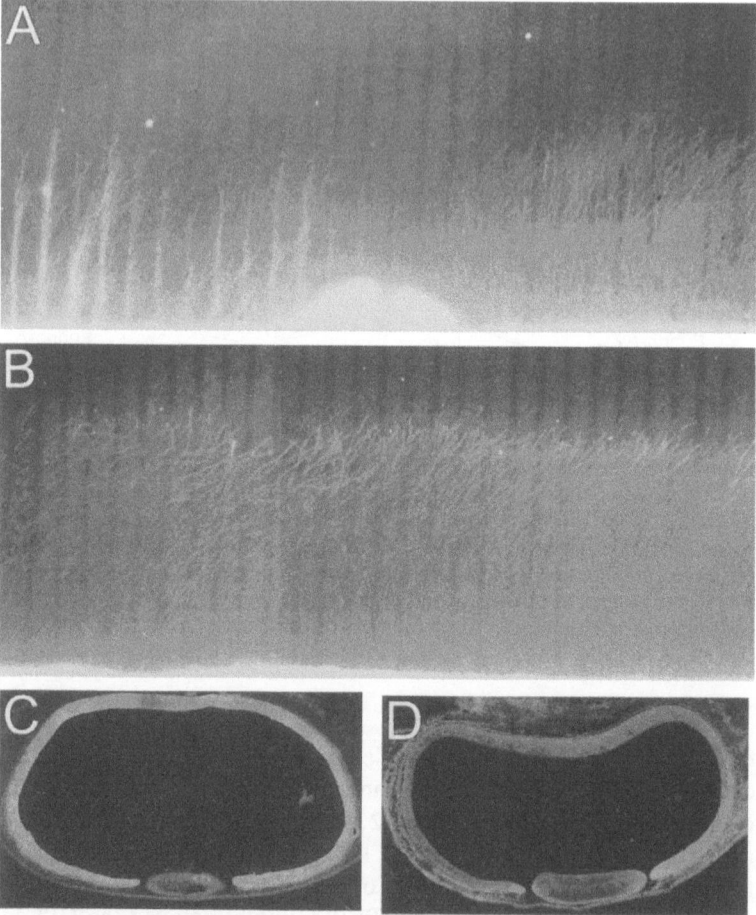

Fig. 2A–D. The striped outgrowth of temporal RGC axons is lost after overexpression of Ephrin-A5. Strips of retinas derived from control-infected (**A**) and Ephrin-A5 (**B**) infected embryos (E7) were analyzed in the stripe assay using alternating rostral and caudal tectal membranes as substrates. Temporal and nasal axons of Ephrin-A5 infected retinas apparently do not discriminate between rostral and caudal membranes (**B**). In controls (**A**) temporal axons, in contrast to nasal axons, showed the typical striped outgrowth. Similar data were obtained after Ephrin-A2 overexpression. Horizontal sections of Ephrin-A5 (**C**) and control-infected (**D**) embryos stained at E4 with a monoclonal antibody against Ephrin-A5 illustrating the level of overexpression usually obtained (nasal part of the retina is oriented to the *right*). Caudal membranes were labeled by addition of fluorescent microspheres, retinal axons were stained by 4-Di-10-Asp (D291, Molecular Probes)

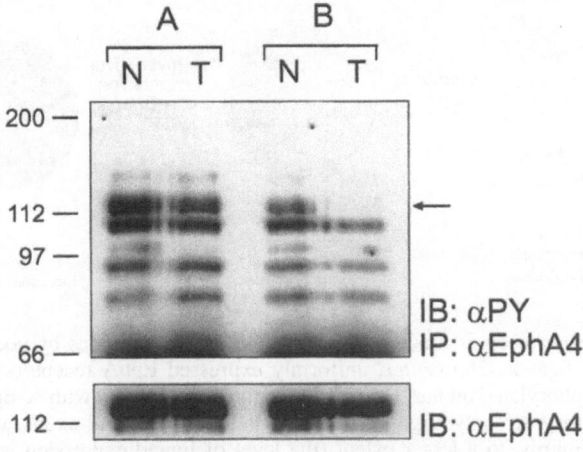

Fig. 3. Tyrosine phosphorylation of the EphA4 receptor in the retina correlates with EphA4/Ephrin-A5 coexpression. *Top:* nasal (*N*) and temporal (*T*) parts of E8 retinas from Ephrin-A5 overexpressing (*A*) and control-infected embryos (*B*) were lysed, the EphA4 receptor was immunoprecipitated (*IP*) by a specific monoclonal antibody, and immunoblotted (*IB*) with the anti-phosphotyrosine antibody 4G10. In Ephrin-A5 overexpressing retinas (*A*) both nasal and temporal receptors are strongly phosphorylated (*arrow*), in control-infected retinas only nasal EphA4 receptors are substantially phosphorylated, in contrast to temporally expressed EphA4 receptors. *Right,* the band corresponding to the EphA4 receptor is indicated by an *arrow. Left,* size marker in kDa. *Bottom,* to verify the immunoprecipitation of similar amounts of EphA4, the filter was stripped and reprobed with an EphA4 antibody

infected with an EphrinA5-expressing retrovirus, the EphA4 receptor on temporal axons is also strongly tyrosine-phosphorylated (Fig. 3).

These data point to an interesting correlation between EphA4 receptor phosphorylation and the behavior of retinal axons in vitro, in that tyrosine phosphorylation of this receptor appears to be linked to an insensitivity of the axons to tectal guidance cues. Temporal axons of control (infected) retinas carrying non-phosphorylated EphA4 receptor show a selective outgrowth in the stripe assay, while nasal axons with a tyrosine-phosphorylated EphA4 show no guidance in stripe assays. In EphrinA5 overexpressing retinas, EphA4 is phosphorylated on both

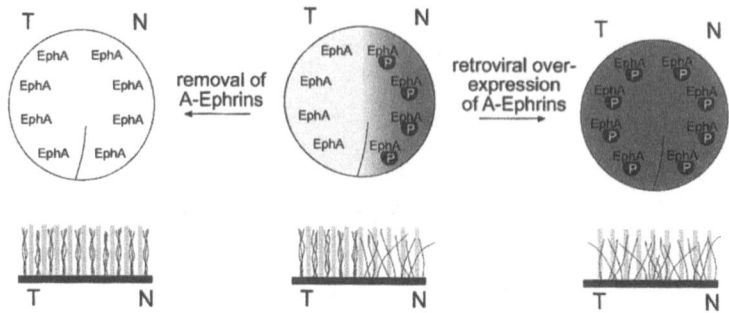

Fig. 4. EphA receptor function is modulated on retinal axons by coexpressed Ephrin-A ligands. *Top center,* uniformly expressed EphA receptors are tyrosine phosphorylated on nasal axons due to the coexpression with A-Ephrins. In contrast, they are only slightly phosphorylated on temporal axons, which express A-Ephrins to a lesser extent (the level of ligand expression is schematized by *gray tones*). *Bottom center,* in the stripe assay, temporal, but not nasal axons are sensitive to Ephrin-A2 resulting in a striped outgrowth. Shedding of Ephrin-A ligands from nasal axons by PI-PLC treatment (*top left*) or overexpression of A-Ephrins on temporal axons (*top right*) changes the behavior of retinal axons in the stripe assay: insensitivity to tectal guidance cues (random outgrowth) correlates with receptor–ligand coexpression and EphA receptor phosphorylation, whereas striped outgrowth, indicating sensitivity to these guidance cues, is seen only for axons with sole EphA receptor expression, i.e., without or with low levels of A-Ephrins (*left*)

temporal and nasal axons and, concomitantly, both types of retinal axon have lost their capacity for selective outgrowth (Fig. 4).

In vivo, retinal overexpression of EphrinA2 led to topographic targeting errors of temporal axons. In a high number of cases temporal axons did not project to their correct retinotopic position on the rostral tectum, but rather to the midtectum and caudal tectum, suggesting a decreased sensitivity of these axons for the repellent activity of the caudal tectum. Thus, data derived from these in vivo experiments were in general agreement with the in vitro data.

In sum, the data suggest that differential ligand expression on retinal axons is a major determinant of topographic targeting in the retinotectal projection (Hornberger et al. 1999).

9.4 Formation of Lamina-Specific Synaptic Connections

Beyond their involvement in the topographic projection of sensory axons, two members of the Eph family, EphrinA5 and EphrinB2, have recently also been implicated in the (non-topographic) processes controlling laminar-specific arborizations (Braisted et al. 1997; Castellani et al. 1998).

Many parts of the vertebrate central nervous system are divided into histologically discrete parallel laminae, onto which distinct populations of afferent axons synapse. These processes appear to be major determinants of specific connectivity in the central nervous system. The molecular and cellular basis of laminar specificity is largely unknown; however, besides other molecules (for review see Sanes and Yamagata 1999) some members of the Eph family have been implicated in these processes. Their activity exclusively in membrane-bound form appears to fit Eph ligands particularly, in contrast to other, soluble guidance molecules, for such a function.

In the chick tectum, retinal axons arborize in three different retinorecipient cell layers, that is, layers B, D, and F of the *stratum griseum et fibrosum superficiale* (SGFS). Here, EphrinB2 is expressed restrictively in layer D and might help guide retinal axons invading the tectum to the appropriate layers. A number of receptors known to bind EphrinB2 are expressed in the retina and presumably on retinal ganglion cell axons. Although its precise function in this patterning process has not been determined, EphrinB2 might function either as an attractant for axons arborizing in layer D or as a repellent for axons projecting into layer B, thus preventing them from growing into deeper tectal layers (Braisted et al. 1997).

EphrinA5, on the other hand, is selectively expressed in layer 4 of the neocortex, whereas one of its receptors, EphA5, is expressed in layers 2/3 and layer 5. Backed up by in vitro studies, the function of EphrinA5 in the cortex might be dual: as a branching factor for layer 6 neurons, which arborize in layer 4, and as a repellent for layer 2/3 neurons, which connect to layer 5 and layers 2/3, thus omitting layer 4. The EphA receptor expressed on layer 6 neurons, possibly mediating the branching activity, has so far not been identified (Castellani et al. 1998).

Here again, a general principle of axon guidance appears to be in operation, i.e., that the function of an axon guidance molecule can be either attractive or repellent, depending on the actual receptor composition found on the interacting neurons.

9.5 Segmental Restriction of Motor Axons, Neural Crest, and Hindbrain Segments

Similar complementary expression patterns of Eph receptors and ligands are seen during pathfinding of motor neuron axons, migration of neural crest cells, and formation of hindbrain segments.

Motor neuron axons leave the neural tube ventrolaterally and extend over the anterior half of each somite, thus avoiding posterior somites expressing EphrinBs. This behavior can be mimicked in vitro in a modified stripe assay resulting in an avoidance by motor neuron axons of stripes containing EphrinB ligands (Wang and Anderson 1997). In hindlimb muscles, EphA4-expressing motor neuron axons avoid regions of EphrinA2 and EphrinA5 expression; in vitro, both ligands have a growth-inhibitory effect on these motor neuron axons (Ohta et al. 1996; Ohta et al. 1997). Thus it appears that during motor neuron pathfinding, both proximally and distally, the Eph family is involved in funneling these axons to their correct target positions.

A similar situation appears to hold true for trunk neural crest cells, which likewise express EphB receptors and migrate through the anterior somites (Krull et al. 1997; Wang and Anderson 1997). In vivo application of function-blocking soluble EphrinB molecules disturbed the segmental pattern of neural crest migration (Krull et al. 1997).

There is complementary expression of EphA4/EphB receptors and EphrinB ligands in hindbrain segments and in the *Xenopus* embryo in streams of branchial neural crest cells migrating to the branchial arches. These receptors and Ephrins appear to mediate repulsion that prevents intermingling of odd- and even-numbered hindbrain segments of branchial neural crest streams (Smith et al. 1997).

9.6 Cardiovascular Development

Blood vessels form de novo from a dispersed population of endothelial cells in a process called vasculogenesis, which occurs in several independent locations during embryonic and extraembryonic development. Subsequently, in angiogenesis, the network of thin tubules is further differentiated into larger, branched vessels through a series of morphogenetic events involving sprouting, splitting and remodelling (Risau and Flamme 1995; Risau 1997). Members of the Eph family have now been implicated in the latter processes (Pandey et al. 1995; Yancopoulos et al. 1998; Adams et al. 1999; Wang et al. 1999).

Initially a clear-cut complementary expression pattern of Eph receptors and ligands was reported, in that EphrinB2 marks arterial, but not venous endothelial cells from the onset of angiogenesis and, conversely, a prominent receptor of EphrinB2, EphB4, marks veins, but not arteries (Wang et al. 1999). Analysis of EphrinB2 knockout mice revealed defects in angiogenesis of both arteries and veins leading apparently to an early lethality, suggesting that Eph-family mediated reciprocal signaling between these two types of vessels is crucial for morphogenesis of the capillary system. However, this more simplistic view has been challenged by another investigation (Adams et al. 1999), which shows a complex involvement of Eph family members, in that a number of Eph receptors and ligands of the B-class are coexpressed on veins and arteries in at least partially overlapping expression patterns. Also, intersomitic vessels that form at somite boundaries express EphB receptors, while EphrinB2 is expressed in the caudal half of the somites. In the EphrinB2 knockout mice, these vessels show an abnormal sprouting behavior (Adams et al. 1999). These initial data suggest a prominent role of the Eph family in angiogenesis.

Interestingly, since Ephrins and Eph receptors have been found to be expressed in a variety of different tumors and tumor cell lines (Wicks et al. 1992; Kiyokawa et al. 1994; Soans et al. 1994), it is possible that these molecules are also involved in the process of neovascularization of growing tumors. With this possibility in mind, it might be worthwhile to investigate whether there is any link between metastasis, on the one hand, and the Eph family controlling cell migration, on the other hand.

Acknowledgements. This work was supported in part by a biotechnology network grant from the European Union and grants from the Deutsche Forschungsgemeinschaft to UD.

References

Adams RH, Wilkinson GA, Weiss C, Diella F, Gale NW, Deutsch U, Risau W, Klein R (1999) Roles of ephrinB ligands and EphB receptors in cardiovascular development: demarcation of arterial/venous domains, vascular morphogenesis and sprouting angiogenesis. Genes Dev 13:295–306

Braisted JE, McLaughlin T, Wang HU, Friedman GC, Anderson DJ, O'Leary DDM (1997) Graded and lamina-specific distributions of ligands of EphB receptor tyrosine kinases in the developing retinotectal system. Dev Biol 191:14–28

Brambilla R, Brückner K, Orioli D, Bergemann AD, Flanagan JG, Klein R (1996) Similarities and differences in the way transmembrane-type ligands interact with the elk subclass of Eph receptors. Mol Cell Neurosci 8:199–209

Brennan C, Monschau B, Lindberg R, Guthrie B, Drescher U, Bonhoeffer F, Holder N (1997) Two Eph receptor tyrosine kinase ligands control axon growth and may be involved in the creation of the retinotectal map in the zebrafish. Development 124:655–664

Castellani V, Yue Y, Gao P-P, Zhou R, Bolz J (1998) Dual action of a ligand for Eph receptor tyrosine kinases on specific populations of axons during the development of cortical circuits. J Neurosci 18:4663–4672

Cheng HJ, Nakamoto M, Bergemann AD, Flanagan JG (1995) Complementary gradients in expression and binding of elf-1 and mek4 in development of the topographic retinotectal projection map. Cell 82:371–381

Connor RJ, Menzel P, Pasquale EB (1998) Expression and tyrosine phosphorylation of Eph receptors suggest multiple mechanisms in patterning of the visual system. Dev Biol 193:21–35

Drescher U, Bonhoeffer F, Müller BK (1997) The Eph family in retinal axon guidance. Curr Opin Neurobiol 7:75–80

Flanagan JG, Leder P (1990) The kit ligand: a cell surface molecule altered in steel mutant fibroblasts. Cell 63:185–194

Flanagan JG, Vanderhaeghen P (1998) The ephrins and Eph receptors in neural development. Annu Rev Neurosci 21:309–45

Frisen J, Yates PA, McLaughlin T, Friedman GC, O'Leary DDM, Barbacid M (1998) Ephrin-A5 (AL-1/RAGS) is essential for proper retinal axon guidance and topographic mapping in the mammalian visual system. Neuron 20:235–43

Gale NW, Holland SJ, Valenzuela DM, Flenniken A, Pan L, Ryan TE, Henkemeyer M, Strebhardt K, Hirai H, Wilkinson DG, Pawson T, Yancopoulos GD (1996) Eph receptors and ligands comprise two major specificity subclasses and are reciprocally compartmentalized during embryogenesis. Neuron 17:9–19

Holt CE, Harris WA (1993) Position, guidance, and mapping in the developing visual system. J Neurobiol 24:1400–1422

Hornberger MR, Dütting D, Ciossek T, Yamada T, Handwerker C, Lang S, Weth F, Huf J, Weßel R, Logan C, Tanaka H, Drescher U (1999) Modulation of EphA receptor function by coexpressed Ephrin-A ligands on retinal ganglion cell axons. Neuron (in press)

Kiyokawa E, Takai S, Tanaka M, Iwase T, Suzuki M, Xiang YY, Naito Y, Yamada K, Sugimura H, Kino I (1994) Overexpression of erk, an eph family receptor protein-tyrosine kinase, in various human tumors. Cancer Res 54:3645–3650

Krull CE, Lansford R, Gale NW, Collazo A, Marcelle C, Yancopoulos GD, Fraser SE, Bronner-Fraser M (1997) Interactions of Eph-related receptors and ligands confer rostrocaudal pattern to trunk neural crest migration. Curr Biol 7:571–580

Marcus RC, Gale NW, Morrison ME, Mason CA, Yancopoulos GD (1996) Eph family receptors and their ligands distribute in opposing gradients in the developing mouse retina. Dev Biol 180:786–789

Mey J, Thanos S (1992) Development of the visual system of the chick – a review. J Hirnforsch. 33:673–702

Monschau B, Kremoser C, Ohta K, Tanaka H, Kaneko T, Yamada T, Handwerker C, Hornberger M, Löschinger J, Pasquale EB, Siever DA, Verderame MF, Müller BK, Bonhoeffer F, Drescher U (1997) Shared and unique functions of RAGS and ELF-1 in guiding retinal axons. EMBO J 16:1258–1267

Müller BK (1999) Growth cone guidance: first steps towards a deeper understanding. Annu Rev Neurosci (in press)

Nakamoto M, Cheng HJ, Friedman GC, McLaughlin T, Hansen MJ, Yoon CH, O'Leary DDM, Flanagan JG (1996) Topographic specific effects of ELF-1 on retinal axon guidance in vitro and retinal axon mapping in vivo. Cell 86:755–766

Ohta K, Iwamasa H, Drescher U, Terasaki H, Tanaka H (1997) The inhibitory effect on neurite outgrowth of motoneurons exerted by the ligands ELF-1 and RAGS. Mech Dev 64:127–135

Ohta K, Nakamura M, Hirokawa K, Tanaka S, Iwama A, Suda T, Ando M, Tanaka H (1996) The receptor tyrosine kinase, Cek8, is transiently expressed on subtypes of motoneurons in the spinal cord during development. Mech Dev 54:59–69

O'Leary DDM, Wilkinson DG (1999) Eph receptors and ephrins in neural development. Curr Opin Neurobiol 9:65–73

Pandey A, Shao HN, Marks RM, Polverini PJ, Dixit VM (1995) Role of B61, the ligand for the eck receptor tyrosine kinase, in TNF-alpha-induced angiogenesis. Science 268:567–569

Pasquale EB (1997) The Eph family of receptors. Curr Opin Cell Biol 9:608–15

Risau W (1997) Mechanisms of angiogenesis. Nature 386:671–674

Risau W, Flamme I (1995) Vasculogenesis. Ann Rev Cell Dev Biol 11:73–91

Sanes JR, Yamagata M (1999) Formation of lamina-specific synaptic connections. Curr Opin Neurobiol 9:79–87

Smith A, Robinson V, Patel K, Wilkinson DG (1997) The EphA4 and EphB1 receptor tyrosine kinases and Ephrin-B2 ligand regulate targeted migration on branchial neural crest cells. Curr Biol 7:561–570

Soans C, Holash JA, Pasquale EB (1994) Characterization of the expression of the cek8 receptor-type tyrosine kinase during development and in tumor-cell lines. Oncogene 9:3353–3361

Sperry RW (1963) Chemoaffinity in the orderly growth of nerve fiber patterns and connections. Proc Natl Acad Sci USA 50:703–710

Stein E, Lane AA, Cerretti DP, Schoecklmann HO, Schroff AD, Van E, Daniel TO (1998) Eph receptors discriminate specific ligand oligomers to determine alternative signaling complexes, attachment, and assembly responses. Genes Dev 12:667–678

Tessier-Lavigne M, Goodman CS (1996) The molecular biology of axon guidance. Science 274:1123–33

Wang HU, Anderson DJ (1997) Eph family transmembrane ligands can mediate repulsive guidance of trunk neural crest migration and motor axon outgrowth. Neuron 18:383–396

Wang HU, Chen Z-F, Anderson DJ (1999) Molecular distinction and angiogenic interaction between embryonic arteries and veins revealed by ephrinB2 and its receptor EphB4. Cell 93:741–753

Wicks IP, Wilkinson D, Salvaris E, Boyd AW (1992) Molecular cloning of HEK, the gene encoding a receptor tyrosine kinase expressed by human lymphoid tumor cell lines. Proc Natl Acad Sci USA 89:1611–1615

Yancopoulos GD, Klagsbrun M, Folkman J (1998) Vasculogenesis, angiogenesis and growth factors: ephrins enter the fray at the border. Cell 93:661–664

Zhou RP (1998) The Eph family receptors and ligands. Pharmacol Ther 77:151–181

10 Embryonic Patterning of Xenopus *Mesoderm by* Bmp-4

C. Niehrs, R. Dosch, and D. Onichtchouk

10.1 Introduction

During embryogenesis three germ layers – ecto-, meso-, and endoderm – give rise to all somatic cells of the vertebrate body. From mesoderm, the skeleton, muscle, connective tissue, heart, endocrine and exocrine organs, blood, and the immune system develop. Mesoderm also plays a critical role during axis formation in the vertebrate embryo. During gastrulation, the early dorsal mesoderm, or Spemann organizer, has a central inductive role. The Spemann organizer transplanted to the ventral side of a host embryo induces a twin embryo containing a complete duplicated embryonic axis including the notochord, somites, and central and peripheral nervous system. The organizer is also present in all other vertebrates analyzed where it elicits comparable effects. In chicken and mice this structure is the node, in fish the embryonic shield, which when transplanted induces ectopic axial tissue (Storey et al. 1992; Beddington 1994; Shih and Fraser 1996). The organizer dorsalizes surrounding

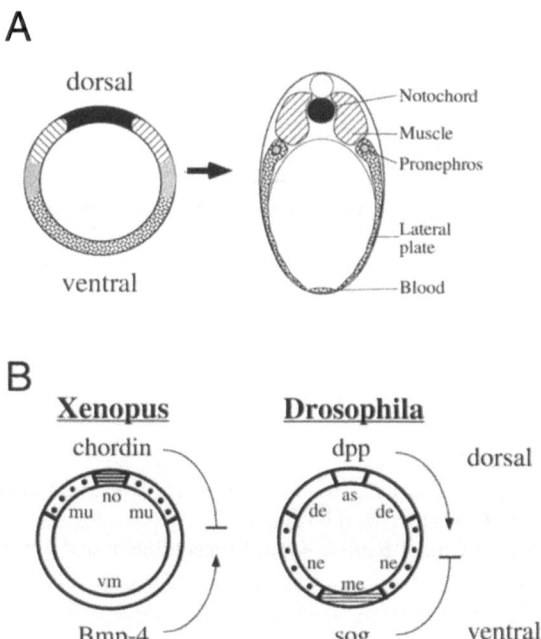

Fig. 1. A Development of *Xenopus* mesoderm. Mesoderm develops from an equatorial belt of cells located between the animal and vegetal pole of the gastrula embryo, the marginal zone (*left*). The marginal zone is already specified dorsoventrally and will give rise to a characteristic sequence of mesodermal tissues shown for the tadpole embryo (*right*). **B** Evolutionary conservation of dorsoventral (d/v) patterning in *Xenopus* and *Drosophila*. The expression of *Xenopus chordin* and *Drosophila short gastrulation* (*sog*) is reversed with respect to the d/v axis as is expression of *Bmp-4* and its *Drosophila* homolog *decapentaplegic* (*dpp*). These factors antagonize each other leading to d/v patterning of *Xenopus* mesoderm into notochord (*no*), muscle (*mu*), and ventral mesoderm (*vm*), as well as *Drosophila* blastoderm into amnioserosa (*as*), dorsal ectoderm (*de*), neural tissue (*ne*), and mesoderm (*me*)

mesoderm, induces and patterns the neural plate, and sets in motion a cascade of secondary inductions (Hamburger 1988; Harland and Gerhart 1997).

In the early amphibian embryo, mesoderm originates from the so-called marginal zone, a ring of cells present in the equatorial region of the early gastrula. The marginal zone arises as the consequence of an inductive process, mesoderm induction (Nieuwkoop 1969; Nieuwkoop 1973; Harland and Gerhart 1997). Based on self-differentiation capacity, four different areas of the marginal zone can be distinguished: the dorsal domain, which is equivalent to the Spemann organizer, differentiates into notochord; the dorsolateral domain into muscle; the lateral and ventrolateral domain into pronephros and mesenchyme; and the ventral domain into blood and mesenchyme (Fig. 1A) (Dale and Slack 1987). Thus, the dorsoventral (d/v) pattern of these tissues is already specified in the gastrula marginal zone. Hence, the elucidation of the molecular mechanisms underlying pattern formation in the marginal zone is of prime importance for understanding mesodermal development.

Bone morphogenetic proteins (BMPs) play an important role in mesodermal patterning. They are members of the TGF-β growth factor superfamily. BMPs can form homo- and heterodimers which exhibit distinct activities (Aono et al. 1995; Suzuki et al. 1997b; Nishimatsu and Thomsen 1998), and they are involved in a variety of developmental processes (Hogan 1996). Here we will discuss progress made in understanding the role of BMP signalling during mesodermal patterning of the amphibian gastrula. The role of BMP signalling in early zebrafish is discussed by M. Hild et al. in this volume (see Chap. 6).

10.2 An Evolutionarily Conserved System for Dorsoventral Patterning

Grafting experiments have shown that during gastrulation, signals emanating from the Spemann organizer dorsalize the ventral marginal zone to form intermediate types of mesoderm (for review see Kimelman et al. 1992; Slack 1993). Ventral mesoderm, therefore, used to be considered ground state mesodermal tissue which serves as the passive substrate upon which the organizer acts.

A number of studies have suggested that the ventral marginal zone not only requires active signals for the specification of the ventral state, but signals to antagonize the organizer. Firstly, the ventral marginal zone expresses peptide growth factors, *Xwnt-8* and *bone morphogenetic proteins -4* and *-7* (*Bmp-4/-7*), that are able to override dorsal mesodermal specification (Köster et al. 1991; Dale et al. 1992; Jones et al. 1992; Christian and Moon 1993; Fainsod et al. 1994; Hemmati-Brivanlou and Thomsen 1995; Schmidt et al. 1995b; Wang et al. 1997). Secondly, inhibition of BMP signalling in *Xenopus* by microinjection of mRNA coding for a dominant-negative BMP receptor (Maeno et al. 1994; Suzuki et al. 1994; Ishikawa et al. 1995), *Bmp-4* antisense mRNA (Steinbeisser et al. 1995), or dominant-negative BMPs (Hawley et al. 1995; Nishimatsu and Thomsen 1998) leads to dorsalization of ventral mesoderm. These findings suggested that marginal zone patterning may be the result of antagonizing dorsal and ventral signals (Sive 1993; Harland 1994; De Robertis and Sasai 1996).

Indeed, the organizer secretes three dorsalizing signals, Noggin (Smith and Harland 1992; Smith et al. 1993) Chordin (Sasai et al. 1994; Sasai et al. 1995), and Follistatin (Hemmati-Brivanlou et al. 1994). All are expressed in the organizer and, when overexpressed in ventral mesoderm, lead to dorsalization. These proteins do not seem to transmit signals through their own receptors. Instead, they directly bind to BMP-2 and BMP-4, thereby inactivating them (Piccolo et al. 1996; Zimmerman et al. 1996; Fainsod et al. 1997; Iemura et al. 1998). Thus, one mechanism for dorsalization is to antagonize ventralizing BMP signals in the marginal zone.

The role of BMPs and its antagonists appears to be evolutionarily conserved between arthropods and vertebrates (Fig. 1B). *Chordin* is homologous to the *Drosophila* gene *short gastrulation* (*sog*), which plays an important role in the d/v development of the fly blastoderm. In *Drosophila*, *sog* antagonizes the *Bmp-4* homolog *decapentaplegic* (*dpp*) (Francois et al. 1994), and overexpression of *Drosophila sog* in *Xenopus* has effects similar to overexpression of *chordin* (Holley et al. 1995; Schmidt et al. 1995a). The astacin protease *tolloid* is involved both in *Drosophila* and in *Xenopus* in antagonizing chordin/sog function by cleaving chordin/BMP complexes (Marqués et al. 1997; Piccolo et al. 1997). *Dpp* functions as a morphogen, and regulates in a dose-dependent fashion the development of ectodermal structures in the fly embryo

(Ferguson and Anderson 1992; Wharton et al. 1993). The discovery of the homology between the antagonizing systems *chordin/sog* and *Bmp-4/dpp* in *Xenopus* and *Drosophila* strongly argues in favor of the thesis of Geoffroy St. Hilaire, that vertebrates and arthropodes have homologous body plans, and that during evolution an inversion of the d/v axes must have taken place (Arendt and Nübler-Jung 1994; De Robertis and Sasai 1996).

Similar to *dpp* regulation of d/v pattern formation in both *Drosophila* ectoderm (Ferguson and Anderson 1992; Wharton et al. 1993) and mesoderm (Wilson et al. 1993; Staehling-Hampton et al. 1995), it seems that in the frog the *Bmp-4/chordin* system regulates d/v pattern formation of all three germ layers. In ectoderm, *chordin* induces and *Bmp-4* inhibits neural tissue formation and instead induces epidermis (Sasai et al. 1995; Wilson and Hemmati-Brivanlou 1995); in endoderm *Bmp-4* represses and *chordin* induces dorsal endoderm (Sasai et al. 1996). Indeed, all known BMP antagonists are able to neuralize naive ectoderm, suggesting that the ectodermal ground state is neural (Hemmati-Brivanlou and Melton 1997).

10.3 *Bmp-4* Acts as a Morphogen

The conservation of d/v patterning in arthropods and vertebrates has raised the possibility that like *dpp* in *Drosophila* (Ferguson and Anderson 1992; Wharton et al. 1993), *Bmp-4* acts as a morphogen in vertebrates (Ferguson 1996; Hogan 1996; Holley et al. 1996; Piccolo et al. 1996; Zimmerman et al. 1996). Support for the possibility that a TGF-β type growth factor acts as a morphogen in vertebrate mesoderm patterning comes from the observation that increasing concentrations of activin and *Vg1* induce a ventrodorsal sequence of mesoderm from uninduced animal cap cells (Green and Smith 1990; Green et al. 1992; Gurdon et al. 1994; Kessler and Melton 1994). However, while this strongly supports the notion of a morphogen functioning in d/v patterning, there is neither evidence for a graded requirement nor evidence for a graded distribution of an activin-like molecule at present. Furthermore, there is recent evidence that both activin and *Vg1* pattern mesoderm indirectly through a relay mechanism (Reilly and Melton 1996). This has left the question of the natural morphogen unresolved.

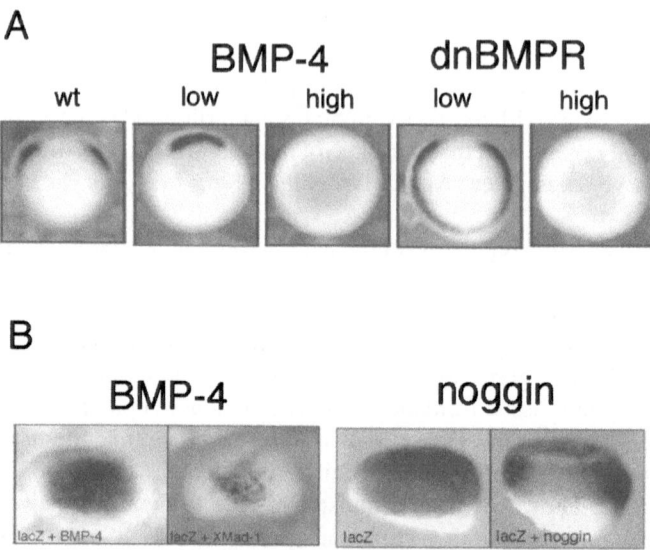

Fig. 2. A Dose-dependent regulation of *myf-5* by BMP-4 signalling. *myf-5* expression occurs in two dorsolateral stripes in the marginal zone of the *Xenopus* gastrula, shown in vegetal view, dorsal side up (*wt*). Ectopic expression of *Bmp-4* or dominant-negative BMP receptor (dnBMPR) mRNAs regulate the pattern of *myf-5* expression at different doses (*low, high*), indicating that stripe formation occurs within a window of the BMP-4 morphogen gradient. **B** Direct and long-range action of BMP-4 and Noggin. *Left:* embryos were microinjected with mRNA encoding β-galactosidase (*lacZ*) showing nuclear localization and either *Bmp-4* or *Xsmad1* (XMAD-1) as indicated, followed by treatment with LiCl to inhibit endogenous BMP target gene expression. Expression of induced *Xvent-1* was detected by whole-mount in situ hybridization in gastrulae. Note that *Xvent-1* expression extends beyond the injected cells (*labeled nuclei*) in the *Bmp-4*-injected embryo, while it is restricted to the β-galactosidase positive cells in *Xsmad1*-injected embryos. Embryos are shown in lateral view with the animal side up. *Right*, embryos were microinjected with β-galactosidase lineage tracer mRNA with or without *noggin* mRNA as indicated. Expression of induced *Xvent-2* was detected by whole-mount in situ hybridization in gastrulae. Note the halo of *Xvent-2* repression beyond the injected cells (labeled nuclei) in the *noggin*-injected embryos. (From Dosch et al. 1997)

If *Bmp-4* acts as a vertebrate morphogen for mesodermal patterning it should fulfill certain criteria in addition to being expressed at the right time and place: (a) it should elicit dose-dependent effects, (b) there should be a graded requirement for *Bmp-4*, (c) its activity should be graded in vivo, and (d) indirect (relay) effects should be excluded.

We have investigated the possibility that *Bmp-4* functions dose-dependently in d/v patterning of *Xenopus* mesoderm at the gastrula stage (Dosch et al. 1997). In marginal zone explants, *Bmp-4* ventralizes dorsal mesoderm in a dose-dependent manner, from notochord, muscle, pronephros to blood. Inversely, a dominant-negative BMP receptor dorsalizes ventral mesoderm dose-dependently from blood, pronephros, muscle to notochord. This dose-dependent effect of BMP signalling can be already observed at the gastrula stage using marker gene expression. Most instructive is the response of the myogenic gene *myf-5*, expressed in two dorsolateral stripes corresponding to prospective muscle. These stripes are the result of a response window corresponding to low BMP signalling, since either lower or higher BMP signalling shuts off *myf-5* expression (Fig. 2A). Thus, for muscle fate a particular threshold of BMP signalling is required (Dosch et al. 1997; Jones and Smith 1998). These results, as well as the response of other marker genes, indicate that different *Bmp-4* doses are both necessary and sufficient for patterning of at least three domains in the gastrula marginal zone, as well as for terminal differentiation into four mesodermal tadpole tissues. There are possibly additional domains in the gastrula marginal zone, e.g., a lateral domain corresponding to future pronephric mesoderm. Recently, a novel ventral domain was discovered marked by the expression of *sizzled* (Salic et al. 1997).

Is there graded BMP-4 signalling in vivo? Although *Bmp-4* and *Bmp-7* are expressed at the right time and place to pattern mesoderm in the early gastrula, their mRNAs do not appear to be distributed in a graded fashion in the marginal zone (Fainsod et al. 1994; Hawley et al. 1995; Schmidt et al. 1995b; our unpublished data). Instead, BMP activity, rather than protein, appears to be graded due to antagonizing action of *noggin, chordin,* and *follistatin* (Ferguson 1996; Hogan 1996; Holley et al. 1996; Piccolo et al. 1996; Zimmerman et al. 1996). Indeed, *Xenopus noggin* has dose-dependent effects on mesodermal (ReemKalma et al. 1995; Dosch et al. 1997; Jones and Smith 1998) and neural patterning (Knecht and Harland 1997; Wilson et al. 1997; Marchant et al. 1998), as would be expected if the ratio between BMP-4 and BMP antagonists determines cell fates.

While dose-dependent patterning by BMP-4 would be expected if the protein acts as a morphogen, the possibility remains that BMP-4 only induces different doses of a yet unknown morphogen. This raises the question of whether BMP-4 protein is able to signal directly over a long range as dpp is in *Drosophila* (Lecuit et al. 1996; Nellen et al. 1996). To address this question, *Bmp-4* mRNA was microinjected into blastomeres of 32-cell stage embryos together with β-galactosidase containing a nuclear localization signal as lineage tracer. Embryos were subsequently dorsalized by incubation in LiCl to inhibit ventral gene expression in order to visualize an induced marker gene, *Xvent-1*. Figure 2B shows that *Xvent-1* is induced by *Bmp-4* and that the expression extends approximately five to ten cell diameters beyond the injected cells containing light-labeled nuclei (lacZ), indicating that the induction of *Xvent-1* by *Bmp-4* is non-cell-autonomous and long-range.

However, the possibility remains that *Bmp-4* acts by inducing some other factor that is ultimately responsible for long-range signalling. To address this possibility, we have made use of *Xsmad1*, an intracellular downstream target of BMP-4 signalling, that mediates its effects, presumably by directly interacting with *Bmp-4* target genes. Like *Bmp-4*, *Xsmad1* induces ventral mesoderm and ventralizes dorsal mesoderm. It rescues the effects of dominant-negative BMP receptor on mesoderm as well as ectoderm (Sekelsky et al. 1995; Graff et al. 1996; Hoodless et al. 1996; Liu et al. 1996; Meersman et al. 1997; Niehrs 1996). If *Bmp-4* acts indirectly by inducing another signalling factor, then *Xsmad1*, faithfully reproducing all effects of *Bmp-4*, should also function at long range, despite being a transcription factor. If, however, *Bmp-4* acts directly on mesodermal cells, only expression of the cytokine, but not that of its intracellular transducer, should yield long-range effects. To distinguish between these possibilities *Xsmad1* mRNA was coinjected with β-galactosidase as lineage tracer and embryos were dorsalized by incubation in LiCl. Figure 2B shows that *Xvent-1* is induced by *Xsmad1*. Unlike in the case of *Bmp-4* injection, expression of *Xvent-1* is restricted to the β-galactosidase positive cells, indicating that the induction of *Xvent-1* by *Xsmad1* is cell-autonomous. We cannot exclude the possibility, however, that BMP-4 signalling involves a relay that is independent of *Xsmad1* function, e.g., by some novel receptor. Given that BMP-4 is auto-activating (Dale et al. 1992), possibly by an autoregulatory loop involving *Xvent-2* (Onichtchouk et al. 1996; Schmidt et al. 1996), it is puzzling that the protein should not be acting by a relay with itself. Clearly, we still know little about the range of BMP-4 action in vivo.

Like BMP-4, its antagonist Noggin has long-range effects (Fig. 2B). When *noggin* mRNA is coinjected with β-galactosidase and the expression of the *Xvent-2* marker gene analyzed, repression of *Xvent* gene expression occurs in a halo around the β-galactosidase-positive cells. This effect can sweep over most of the embryo at higher doses, suggesting that Noggin can diffuse much further than BMP-4 (Dosch et al. 1997; Jones and Smith 1998).

These results argue for a model (Fig. 3), where positional information in the gastrula marginal zone is provided by graded BMP-4 activity which is high ventrally and low dorsally. Graded BMP-4 activity is the result of superimposition of antagonizing BMP-4 and BMP antagonists, e.g., Noggin, Chordin, and Follistatin. In the dorsolateral domain BMP-4 protein is attenuated, allowing for *myf-5* expression and muscle differentiation, both of which require BMP signalling. Yet, muscle differentiation and *myf-5* expression are repressed by high BMP-4 levels, which induce blood differentiation and ventral marker gene (*Xvent-1*) expression. Thus, dorsal, dorsolateral, and lateroventral fate is determined by distinct BMP-4 concentrations. Although the BMP-4 activity gradient remains to be visualized by some direct means, both Noggin and BMP-4 are capable of signalling at long range in the marginal zone as would be expected in this model. Interestingly, in the animal cap BMP-4 appears to have short-range action (Jones et al. 1996). This suggests regional differences in diffusibility of BMP-4. The shape of the BMP-4 activity gradient may therefore be influenced by these differences, possibly mediated by the extracellular matrix.

The BMP antagonists Noggin, Chordin, and Follistatin have all the properties of a bona fide inducer of dorsal and dorsolateral fate. Given the observation, however, that these BMP antagonists function by sequestering and neutralizing BMP proteins (Piccolo et al. 1996; Zimmerman et al. 1996; Fainsod et al. 1997; Iemura et al. 1998), they may not be considered mechanistically instructive inducers unless a separate receptor is identified. Rather, they modify the positional information which is provided by BMP signalling, thus generating graded activity of the BMP-4 morphogen. In this mechanistic view, the organizer may

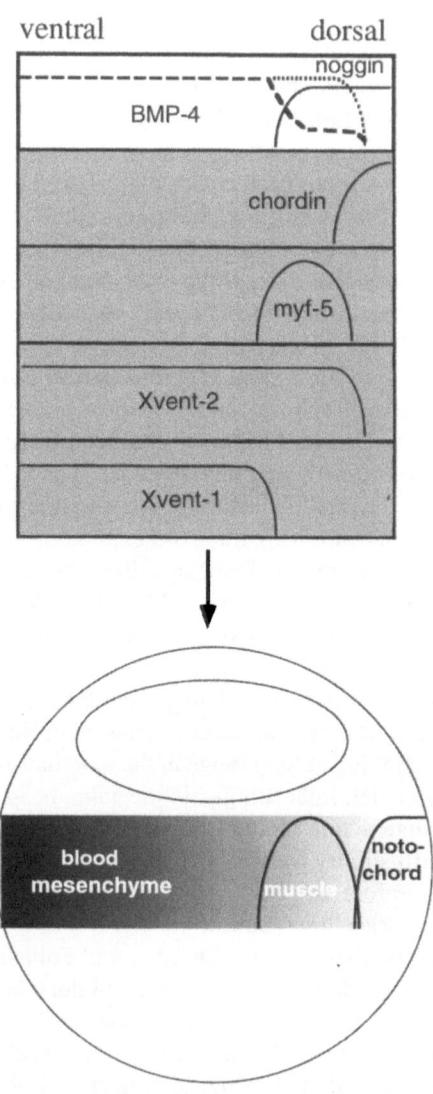

Fig. 3. Legend see p. 175

have a passive role in signalling, because positional information read by cells in the marginal zone corresponds to the local BMP-4 activity, which becomes graded by organizer signals. Put differently, what appears macroscopically to be induction (dorsalization), is microscopically modification of ventral morphogen read-out.

While we have focused here on *Bmp-4*, there are at least four other TGF-β type BMPs expressed in early *Xenopus* embryos: *Bmp-2* (Nishimatsu et al. 1992; Clement et al. 1995; Hemmati-Brivanlou and Thomsen 1995), two *Bmp-7* related genes (Hawley et al. 1995; Wang et al. 1997), and anti-dorsalizing morphogenetic protein (ADMP) (Moos et al. 1995), all of which are ventralizing. While maternally expressed *Bmp-2* may be involved in mesoderm induction, the low levels of protein present at gastrulation and the fact that antisense injection of *Bmp-2* mRNA does not show neuralizing effects [in contrast to antisense *Bmp-4* mRNA; (Sasai et al. 1995)], makes a function during gastrulation less likely. *Bmp-7* is expressed in the marginal zone during gastrula stages and shows activities similar to *Bmp-4*, suggesting that it may act in concert with *Bmp-4*. In addition, it can form functional heterodimers with *Bmp-4* (Nishimatsu and Thomsen 1998). ADMP is, curiously, expressed in the organizer and may be important to fine-tune coexpressed BMP antagonists (Moos et al. 1995).

Disruption of the mouse *Bmp-4* gene is embryonically lethal, with embryos dying around gastrulation, but penetrance is variable (Winnier et al. 1995). In contrast, *Bmp-7* (Dudley et al. 1995) and *Bmp-2* (Zhang and Bradley 1996) mouse mutants do not show defects in d/v patterning. This suggests that in mice, various BMPs act redundantly during gastru-

◄───────────────────────────────────────

Fig. 3. Model of mesodermal patterning at the gastrula stage by a *Bmp-4* morphogen gradient. *Bmp-4* mRNA is uniformly expressed in the marginal zone except in the dorsal domain (*fine dashed line*). *Noggin* (*continuous line*) is expressed in the dorsal domain. Their protein products overlap dorsolaterally leading to attenuated BMP-4 activity (*coarse dashed line*), resulting in dorsal (future notochord), dorsolateral (future muscle), and lateroventral (future blood, lateral plate mesenchyme) positional values. *Xvent-1* has a high dose requirement for BMP-4. *myf-5* has a requirement for a low ,and is inhibited by a high, BMP-4 dose. *Xvent-2* has a requirement for a low BMP-4 dose. The *black and white gradient* in the *lower panel* indicates the BMP-4 activity gradient in the gastrula marginal zone which regulates formation of the indicated mesodermal tissues. (From Dosch et al. 1997)

lation. In the zebrafish mutant swirl, a *bmp-2* gene is mutated and embryos show expanded dorsolateral structures at the expense of ventral ones, both in ectoderm and mesoderm (Kishimoto et al. 1997; Nguyen et al. 1998). The difference in severity between mouse and fish *Bmp-2* mutants may be explained by the fact that swirl is a dominant-negative mutant, which probably affects the function of redundant BMPs, e.g., *Bmp-4* (Kishimoto et al. 1997).

In *Xenopus* a *Bmp-4* (and *Bmp-7*) morphogen system seems also involved in early neurogenesis, where *Bmp-4* inhibits, and *chordin*, *noggin*, and *follistatin* promote neuralization (Lamb et al. 1993; Hemmati-Brivanlou et al. 1994; Hawley et al. 1995; Sasai et al. 1995; Wilson and Hemmati-Brivanlou 1995; Suzuki et al. 1997c). Like in mesodermal patterning BMP-4 acts in a concentration-dependent manner in neural induction. With decreasing doses three different ectodermal structures, epidermis, neural crest/cement gland, and anterior neural tissue, are induced (Knecht and Harland 1997; Wilson et al. 1997; Marchant et al. 1998). Similar results are observed in zebrafish (Kishimoto et al. 1997; Neave et al. 1997; Nguyen et al. 1998). Finally, d/v patterning of endoderm is also regulated by BMP-4 (Sasai et al. 1996). Thus, the BMP-4 morphogen gradient patterns all three germ layers (De Robertis and Sasai 1996).

10.4 Downstream of *Bmp-4*:
Xvent Homeoproteins and Smads

An important question is how the quantitative differences in BMP-4 signalling are converted into qualitative discrete cellular responses. BMP proteins bind to type I (Graff et al. 1994; Suzuki et al. 1994) and type II BMP (Frisch and Wright 1998) transmembrane receptor serine/threonine kinases which form oligomers upon ligand binding. Following receptor activation, BMP signalling is transduced by members of the Smad family of DNA binding proteins (Massague et al. 1997). There are three types of Smads: receptor activated Smads, co-Smads, and anti-Smads. Receptor activated *Smad-1* and *-5* act as specific substrates for ligand-activated type II BMP receptor kinases (Thomsen 1996; Meersman et al. 1997; Suzuki et al. 1997a). The co-Smad-4 forms a complex with receptor activated Smad-1 and -5 upon their phosphorylation which is required for transcriptional activation of Smad target genes (Chen et al. 1997; Liu et al. 1997; Zhang et al. 1997). The BMP-specific Smad-1 may compete with the activin-specific Smad-2 for binding of Smad-4, suggesting an intracellular antagonism between the two signalling pathways (Candia et al. 1997). Finally, anti-Smad-6 and -7 serve to inhibit BMP signalling and seem to be involved in negative feedback regulation (Casellas and Hemmati-Brivanlou 1998; Hata et al. 1998).

The disruption of Smad function is the basis of a number of cancers in man (Massague et al. 1997). *Smad-4* is a tumor suppressor gene which is disrupted in a large proportion of pancreatic carcinomas. *Smad-2* disruption correlates with head and neck carcinomas. Targeted disruption of *Smad-3* in mice leads to colon carcinoma (Zhu et al. 1998). Together with the observation that BMPs are expressed in many adult organs, this suggests that components of the BMP pathway are prime candidates for genes implicated in a variety of cancers.

Interestingly, receptor activated Smads are able to act in a dose-dependent fashion (Graff et al. 1996; Wilson et al. 1997). Thus, the conversion of positional information into qualitatively distinct responses is likely to occur at the level of induced target genes which mediate BMP signalling. In the *Drosophila* wing, the transcription factors *spalt*, *spalt-related* and *optomotor blind* (*omb*) are expressed in nested domains whose boundaries of expression are a function of the distance from a local Dpp source (Lecuit et al. 1996; Nellen et al. 1996). These genes have important roles in mediating the transcriptional effects at distinct concentration thresholds downstream of the *dpp* morphogen (De Celis et al. 1996; Grimm and Pflugfelder 1996; Sturtevant et al. 1997).

In *Xenopus*, candidate transcriptional targets that mediate the effects of *Bmp-4* are *msx1* in ectoderm (Suzuki et al. 1997) and the *Xvent* homeobox genes in ecto- and mesoderm. *Xvent-1* (Gawantka et al. 1995), also called *PV.1* (Tidman-Ault et al. 1996), *Xvent-2* (Onichtchouk et al. 1996), also called *Xbr1*, and *Vox* and *Xom* (Papalopulu and Kintner 1996; Schmidt et al. 1996; Ladher et al. 1996; see Lemaire 1996 for review), are overlappingly expressed in the gastrula marginal zone but show distinct dorsal boundaries of expression (Fig. 4B,C) (Onichtchouk et al. 1996). Similar to the situation in *Drosophila* where the anterior expression boundaries of *omb* and *spalt* are regulated by the distance from the *dpp* source, the different dorsal

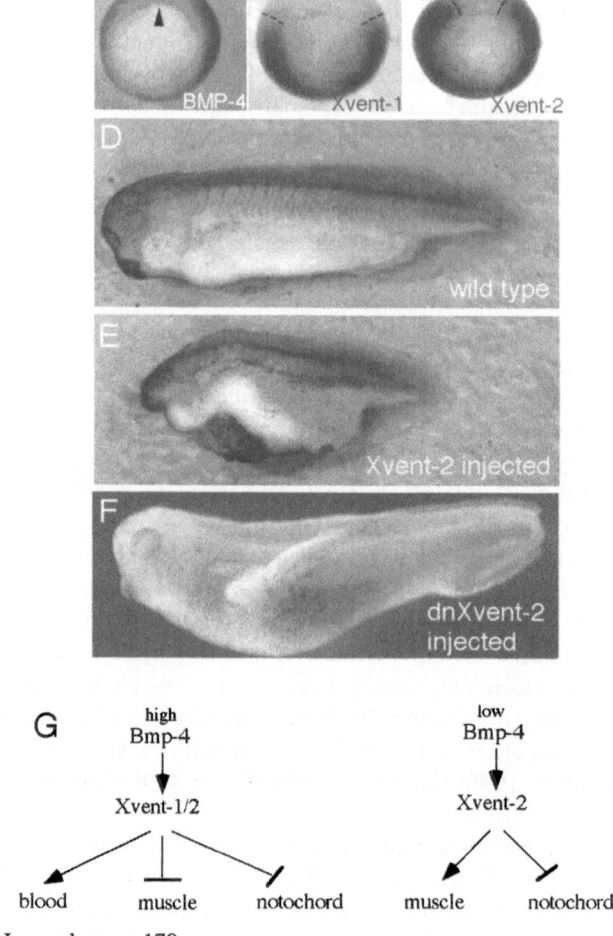

Fig. 4. Legend see p. 179

boundaries of *Xvent-1* and *Xvent-2* expression are regulated by the local activity of BMP-4 in the marginal zone (Fig. 3). *Xvent-2* requires a lower BMP-4 dose to be induced than *Xvent-1*, and hence *Xvent-2* is expressed closer to the organizer, where BMP signalling is attenuated, compared to *Xvent-1* (Dosch et al. 1997).

Consistent with *Xvent* genes acting in mediating BMP signalling, overexpression on the dorsal side of both *Xvent* genes leads to ventralization of embryos (Fig. 4E) (Gawantka et al. 1995; Ladher et al. 1996; Onichtchouk et al. 1996; Schmidt et al. 1996; Tidman-Ault et al. 1996). Xvent proteins function as transcriptional repressors and dominant-negative forms have been generated by fusing their homeodomains to the transcriptional activation domain of VP16 (Onichtchouk et al. 1998). Microinjection of synthetic mRNA encoding dominant-negative *Xvent* mRNAs leads to dorsalization of ventral mesoderm and induction of secondary embryonic axes (Fig. 4F). The organizer gene *XFD-1'* is repressed by, and is a direct target gene of, *Xvent-1* (Friedle et al. 1998). The DNA target consensus sequence recognized by *Xvent-1* is CTATTTG, and thus is distinct from typical homeodomain consensus sequences, probably due to the presence of threonine instead of the more common valine or isoleucine residues in the DNA recognition helix which is conserved in the Xvent protein family.

◄ ───

Fig. 4A–G. *Xvent* homeobox genes mediate BMP signalling. Whole-mount in situ hybridizations showing expression of *Bmp-4* (**A**), *Xvent-1* (**B**), and *Xvent-2* (**C**) in *Xenopus* gastrulae, vegetal view dorsal side up (*arrowhead* indicates the dorsal blastopore lip). Note the different dorsal expression boundaries of *Xvent-1* and *Xvent-2* (*dashed lines*) indicative of different threshold responses for BMP signalling. **D** Control embryo. **E** Embryo microinjected with *Xvent-2* mRNA showing loss of head and axial structures, indicative of ventralization. **F** Embryo microinjected with dominant negative (*dn*) *Xvent-2* mRNA showing a secondary embryonic axis, indicative of dorsalization. **G** Model of *Xvent* gene action in response to different BMP-4 concentrations. *Left*, high BMP-4 activity in lateroventral marginal zone drives the expression of both *Xvent* genes. Coexpression of both *Xvent* genes specifies cells as the most ventral type of mesoderm: blood and mesenchyme, repressing muscle and notochord. *Right*, low BMP-4 activity in dorsolateral marginal zone is not sufficient for *Xvent-1* but for *Xvent-2* expression. *Xvent-2* expression promotes muscle differentiation, but represses notochord. (From Onichtchouk et al. 1996; Onichtchouk et al. 1998)

Xvent-1 and Xvent-2 share many functional features but what distinguishes them is their effect following wild-type overexpression, where Xvent-2 is able to ventralize embryos in a dose-dependent manner, while Xvent-1 causes milder ventralization and lethality at a higher dose (Gawantka et al. 1995; Onichtchouk et al. 1996). In addition, unlike VPXvent-2, VPXvent-1 induces secondary embryonic axes that frequently contain notochords (Onichtchouk et al. 1998). Yet, Xvents have very similar DNA binding domains and in both gain- and loss-of-function experiments affect expression of a similar set of marker genes suggesting that they share the same targets. They possibly recruit distinct cofactors via their unique N-terminal domains.

The functional similarities taken together with the distinct d/v expression domains of Xvent-1 and -2 support a scenario where the sum of Xvent activity determines the specification of mesodermal cells (Fig. 4G). The dorsolateral boundaries of Xvent gene expression are determined by the local activity of BMP-4 (Dosch et al. 1997). In ventrolateral mesoderm both Xvents are expressed and the sum of Xvent activity is high, suppressing muscle and notochord and allowing mesenchyme and blood differentiation. In dorsolateral mesoderm, only Xvent-2 is expressed which suppresses notochord but allows muscle differentiation. Firstly, this model is supported by the finding, that Xvents act in an additive fashion both in gain- and loss-of-function experiments. Secondly, a high dose of Bmp-4 mRNA injection induces both Xvent genes and represses muscle and notochord differentiation, while a low dose of Bmp-4 induces Xvent-2 and muscle and represses notochord differentiation (Dosch et al. 1997).

Figure 5 shows a diagram of the BMP signal transduction pathway. It is interesting that a variety of components of the pathway appear to be regulated by BMP signalling itself, since they show the characteristic complex expression pattern of Bmp-4 at tadpole stage (Gawantka et al. 1998; Hata et al. 1998; Nakayama et al. 1998a; Takase et al. 1998; Bhushan et al. 1998; Casellas and Hemmati-Brivanlou 1998; Nakayama et al. 1998b). Indeed, there are positive feedback loops between Bmp-4 and Xvent-2 in Xenopus (Onichtchouk et al. 1996; Schmidt et al. 1996), as well as bmp-4 and bmp-2 in zebrafish (Kishimoto et al. 1997; Nguyen et al. 1998). The diagnostic expression pattern of Bmp-4 may therefore be useful to screen for novel components of the pathway as was the case for Xvent-2 (Onichtchouk et al. 1996). We have also observed the

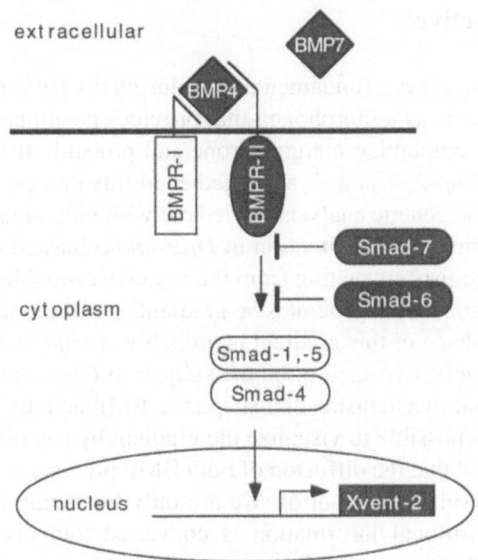

Fig. 5. The bone morphogenetic protein *(BMP)* signal transduction pathway and synexpression group. Members of the BMP growth factors bind to and activate type I and type II receptor serine/threonine kinases leading to phosphorylation and activation of BMP receptor-specific Smad-1 and -5. Upon phosphorylation Smad-1 and -5 bind to a co-Smad (Smad-4) and enter the nucleus where they associate with yet unknown transcription factors and activate transcription of target genes such as *Xvent-2*. Inhibitory Smad-6 and -7 interfere with signal transduction and limit BMP signalling. Components of the BMP pathway which are coexpressed with, and probably coregulated by, *Bmp-4* are *highlighted* in *gray*. This group of coregulated genes is called the *Bmp-4* synexpression group. *Non-highlighted* components are mostly ubiquitously expressed

phenomenon of coordinated expression of interacting genes in a large-scale screen for three other groups of genes which we refer to collectively as synexpression groups (Gawantka et al., 1998). In the case of the *Bmp-4* synexpression group, coordinated expression may be important for autoregulation of the pathway.

10.5 Perspective

BMP signalling plays a fundamental role during d/v patterning of mesoderm. *Bmp-4* acts as a morphogen that provides positional information
to cells of the amphibian marginal zone and probably the ectoderm. It
appears that *Bmp-2,-4* and *-7* act together in this process which limits
loss-of-function genetic analysis carried out with individual members of
the family. Similar to the situation in *Drosophila* blastoderm, a system
of BMP antagonists emanating from the organizer provides a BMP-sig-
nalling free zone and generates a gradient of BMP activity in the
vicinity. The shape of this gradient is probably crucial for the emerging
pattern and can be expected to be the subject to fine control. We know
very little about mechanisms that shape the BMP activity gradient, nor
has it yet been possible to visualize the gradient by any direct means. It
can be expected that the diffusion of both BMP proteins as well as BMP
antagonists is subject to control. We are only beginning to understand
how BMP positional information is converted into discrete cellular
responses, with *Xvent* homeobox genes being important components in
this process. It will be interesting to identify the promoter elements of
BMP target genes such as *Xvents* or *myf-5* that regulate their distinct
expression domains at different signalling thresholds. Analysis of Smad
function has revealed that components of TGF-β signalling pathways
are prime candidates for cancer genes. Hence, studying the mechanism
of BMP signalling has important implications for understanding cell
differentiation in development as well as in disease.

Acknowledgements. This work was supported by grant Ni 286/4–2 from the
Deutsche Forschungsgemeinschaft.

References

Aono A, Hazama M, Notoya K, Taketomi S, Yamasaki H, Tsukuda R, Sasaki S
and Fujisawa Y (1995) Potent ectopic bone-inducing activity of bone morphogenetic protein 4/7 heterodimer. Biochem Biophys Res Commun 210:
679–677
Arendt D, Nübler-Jung K (1994) Inversion of dorsoventral axis? Nature
371:26

Beddington RS (1994) Induction of a second neural axis by the mouse node. Development 120:613–20

Bhushan A, Chen Y, Vale W (1998) Smad7 inhibits mesoderm formation and promotes neural cell fate in *Xenopus* embryos. Dev Biol 200:260–268

Candia AF, Watabe T, Hawley HB, Onichtchouk D, Zhang Y, Derynck R, Niehrs C, Cho KWY (1997) Cellular interpretation of multiple TGF-beta signals: intracellular antagonism between activin/BVg1 and BMP-2/4 signaling mediated by Smads. Development 124:4467–4480

Casellas R, Hemmati-Brivanlou A (1998) *Xenopus* Smad7 inhibits both the activin and BMP pathways and acts as a neural inducer. Dev Biol 198:1–12

Chen X, Weisberg E, Fridmacher V, Watanabe M, Naco G, Whitman M (1997) Smad4 and FAST-1 in the assembly of activin-responsive factor. Nature 389:85–9

Christian JL, Moon RT (1993) Interactions between Xwnt-8 and Spemann organizer signaling pathways generate dorsoventral pattern in the embryonic mesoderm of *Xenopus*. Genes Dev 7:13–28

Clement JH, Fettes P, Knöchel S, Lef J, Knöchel W (1995) Bone morphogenetic protein 2 in the early development of *Xenopus laevis*. Mech Dev 52:357–370

Dale L, Slack JM (1987) Regional specification within the mesoderm of early embryos of *Xenopus laevis*. Development 100:279–95

Dale L, Howes G, Price BM, Smith JC (1992) Bone morphogenetic protein 4: a ventralizing factor in early *Xenopus* development. Development 115:573–85

De Celis JF, Barrio R, Kafatos FC (1996) A gene complex acting downstream of dpp in *Drosophila* wing morphogenesis. Nature 381:421–424

De Robertis EM, Sasai Y (1996) A common plan for dorsoventral patterning in Bilateria. Nature 380:37–40

Dosch R, Gawantka V, Delius H, Blumenstock C, Niehrs C (1997) Bmp-4 acts as a morphogen in dorsoventral mesoderm patterning in *Xenopus*. Development 124:2325–2334

Dudley AT, Lyons KM, Robertson EJ (1995) A requirement for bone morphogenetic protein-7 during development of the mammalian kidney and eye. Genes Dev 9:2795–807

Fainsod A, Steinbeisser H, De Robertis EM (1994) On the function of BMP-4 in patterning the marginal zone of the *Xenopus* embryo. EMBO J 13:5015–25

Fainsod A, Deißler K, Yelin R, Marom M, Epstein M, Pillemer G, Steinbeisser H, Blum M (1997) The dorsalizing and neural inducing gene follistatin is an antagonist of Bmp-4. Mech Dev 63:39–50

Ferguson EL (1996) Conservation of dorsal-ventral patterning in arthropods and chordates. Curr Opin Genet Dev 6:424–431

Ferguson EL, Anderson KV (1992) Decapentaplegic acts as a morphogen to organize dorsal-ventral pattern in the *Drosophila* embryo. Cell 71:451–461

Francois V, Solloway M, O'Neill JW, Emery J, Bier E (1994) Dorsal-ventral patterning of the *Drosophila* embryo depends on a putative negative growth factor encoded by the short gastrulation gene. Genes Dev 8:2602–16

Friedle H, Rastegar S, Paul H, Kaufmann E, Knöchel W (1998) Xvent-1 mediates BMP-4-induced suppression of the dorsal-lip-specific early response gene XFD-1' in *Xenopus* embryos. EMBO J 17:2298–307

Frisch A, Wright CVE (1998) XBMPRII, a novel *Xenopus* type II receptor mediating BMP signalling in embryonic tissues. Development 125:431–442

Gawantka V, Delius H, Hirschfeld K, Blumenstock C, Niehrs C (1995) Antagonizing the Spemann organizer: role of the homeobox gene Xvent-1. EMBO J 14:6268–79

Gawantka V, Pollet N, Delius H, Pfister R, Vingron M, Nitsch R, Blumenstock C, Niehrs C (1998) Gene expression screening in *Xenopus* identifies molecular pathways, predicts gene function and provides a global view of embryonic patterning. Mech Dev 77:95–141

Graff JM, Thies RS, Song JJ, Celeste AJ, Melton DA (1994) Studies with a *Xenopus* BMP receptor suggest that ventral mesoderm-inducing signals override dorsal signals in vivo. Cell 79:169–79

Graff JM, Bansal A, Melton DA (1996) *Xenopus* Mad proteins transduce distinct subsets of signals for the TGFb superfamily. Cell 85:479–487

Green JB, Smith JC (1990) Graded changes in dose of a *Xenopus* activin A homologue elicit stepwise transitions in embryonic cell fate. Nature 347:391–4

Green JBA, New HV, Smith JC (1992) Responses of embryonic *Xenopus* cells to activin and FGF are separated by multiple dose thresholds and correspond to distinct axes of the mesoderm. Cell 71:731–739

Grimm S, Pflugfelder GO (1996) Control of the gene optomotor-blind in *Drosophila* wing development by decapentaplegic and wingless. Science 271:1601–1604

Gurdon JB, Harger P, Mitchell A, Lemaire P (1994) Activin signalling and response to a morphogen gradient. Nature 371:487–92

Hamburger V (1988). The heritage of experimental embryology. Oxford University Press, New York

Harland RM (1994) The transforming growth factor beta family and induction of the vertebrate mesoderm: bone morphogenetic proteins are ventral inducers. Proc Natl Acad Sci USA 91:10243–6

Harland RM, Gerhart J (1997) Formation and function of Spemann's organizer. Ann Rev Cell Dev Biol 13:611–667

Hata A, Lagna G, Massague J, Hemmati-Brivanlou A (1998) Smad6 inhibits BMP/Smad1 signaling by specifically competing with the Smad4 tumor suppressor. Genes Dev 12:186–97

Hawley SH, Wunnenberg SK, Hashimoto C, Laurent MN, Watabe T, Blumberg BW, Cho KW (1995) Disruption of BMP signals in embryonic *Xenopus* ectoderm leads to direct neural induction. Genes Dev 9:2923–35

Hemmati-Brivanlou A, Kelly OG, Melton DA (1994) Follistatin, an antagonist of activin, is expressed in the Spemann organizer and displays direct neuralizing activity. Cell 77:283–95

Hemmati-Brivanlou A, Thomsen GH (1995) Ventral mesodermal patterning in *Xenopus* embryos: expression patterns and activities of BMP-2 and BMP-4. Dev Genet 17:78–89

Hemmati-Brivanlou A, Melton D (1997) Vertebrate embryonic cells will become nerve cells unless told otherwise. Cell 88:13–17

Hogan BM (1996) Bone morphogenetic proteins: multifunctional regulators of vertebrate development. Genes Dev 10:1580–1594

Holley SA, Jackson PD, Sasai Y, Lu B, De Robertis EM, Hoffmann M, Ferguson EL (1995) A conserved system for dorso-ventral patterning in insects and vertebrates involving sog and chordin. Nature 376:249–253

Holley SA, Neul JL, Attisano L, O'Connor MB, DeRobertis EM, Ferguson EL (1996) The *Xenopus* dorsalizing factor noggin ventralizes *Drosophila* embryos by preventing dpp from activating its receptor. Cell 86:607–617

Hoodless P, Haerry T, Abdollah S, Stapleton M, O'Connor MB, Attisano L, Wrana J (1996) MADR1, a Mad-related protein that functions in BMP2 signaling pathways. Cell 85:489–500

Iemura S-I, Yamamoto T, Takagi C, Uchiyama H, Natsume T, Shimasaki S, Sugino H, Ueno N (1998) Direct binding of follistatin to a complex of bone morphogenetic protein and its receptor inhibits ventral and epidermal cell fates in early *Xenopus* embryos. Proc Natl Acad Sci USA 95:9337–9342

Ishikawa T, Yoshioka H, Ohuchi H, Noji S, Nohno T (1995) Truncated type II receptor for BMP-4 induces secondary axial structures in *Xenopus* embryos. Biochem Biophys Res Commun 216:26–33

Jones CM, Lyons KM, Lapan PM, Wright CV, Hogan BL (1992) DVR-4 (bone morphogenetic protein-4) as a posterior-ventralizing factor in *Xenopus* mesoderm induction. Development 115:639–47

Jones CM, Armes N, Smith JC (1996) Signalling by TGF-beta family members: short range effects of Xnr-2 and BMP-4 contrast with long-range effects of activin. Curr Biol 6:1468–1475

Jones CM, Smith JC (1998) Establishment of a BMP-4 morphogen gradient by long-range inhibition. Dev Biol 194:12–17

Kessler DS, Melton DA (1994) Vertebrate embryonic induction: mesodermal and neural patterning. Science 266:596–604

Kimelman D, Christian JL, Moon RT (1992) Synergistic principles of development: overlapping patterning systems in *Xenopus* mesoderm induction. Development 116:1–9

Kishimoto Y, Lee KH, Zon L, Hammerschmidt M, Schulte MS (1997) The molecular nature of zebrafish swirl: BMP2 function is essential during early dorsoventral patterning. Development 124:4457–66

Knecht A, Harland RM (1997) Mechanisms of dorsal-ventral patterning in noggin-induced neural tissue. Development 124:2477–2488

Köster M, Plessow S, Clement JH, Lorenz A, Tiedemann H, Knöchel W (1991) Bone morphogenetic protein 4 (BMP-4), a member of the TGF-beta family, in early embryos of *Xenopus laevis*: analysis of mesoderm inducing activity. Mech Dev 33:191–9

Ladher R, Mohun TJ, Smith JC, Snape AM (1996) Xom: a *Xenopus* homeobox gene that mediates the early effects of BMP-4. Development 122:2385–2394

Lamb TM, Knecht AK, Smith WC, Stachel SE, Economides AN, Stahl N, Yancopolous GD, Harland RM (1993) Neural induction by the secreted polypeptide noggin. Science 262:713–8

Lecuit T, Brook WJ, Ng M, Calleja M, Sun H, Cohen S (1996) Two distinct mechanisms for long-range patterning by decapentaplegic in the *Drosophila* wing. Nature 381:387–392

Lemaire P (1996) The coming of age of ventralising homeobox genes in amphibian development. Bioessays 18:701–704

Liu F, Pouponnot C, Massague J (1997) Dual role of the Smad4/DPC4 tumor suppressor in TGFbeta-inducible transcriptional complexes. Genes Dev 11:3157–67

Maeno M, Ong RC, Suzuki A, Ueno N, Kung HF (1994) A truncated bone morphogenetic protein 4 receptor alters the fate of ventral mesoderm to dorsal mesoderm: roles of animal pole tissue in the development of ventral mesoderm. Proc Natl Acad Sci USA 91:10260–4

Marchant L, Linker C, Ruiz P, Guerrero N, Mayor R (1998) The inductive properties of mesoderm suggest that the neural crest cells are specified by a BMP gradient. Dev Biol 15:319–329

Marqués G, Musacchio M, Shimmel MJ, Wünnenburg-Stapleton K, Cho KWY, O'Connor MB (1997) Production of a dpp activity gradient in the early *Drosophila* embryo through the opposing actions of SOG and TLD proteins. Cell 91:417–426

Massague J, Hata A, Liu F (1997) TGF-beta signalling through the Smad pathway. Trends Cell Biol 7:187–192

Meersman G, Verschueren K, Nelles L, Blumenstock C, Kraft H, Wuytens G, Remacle J, Kozak CA, Tylzanowsky P, Niehrs C, Huylebroeck D (1997) The C-terminal domain of Mad-like signal transducers is sufficient for biological activity in vivo and transcriptional activation. Mech Dev 61:127–140

Moos MJ, Wang S, Krinks M (1995) Anti-dorsalizing morphogenetic protein is a novel TGF-beta homolog expressed in the Spemann organizer. Development 121:4293–301

Nakayama T, Gardner H, Berg LK, Christian JL (1998a) Smad6 functions as an intracellular antagonist of some TGF-beta family members during *Xenopus* embryogenesis. Genes Cells 3:387–394

Nakayama T, Snyder MA, Grewal SS, Tsuneizumi K, Tabata T, Christian JL (1998b) *Xenopus* Smad8 acts downstream of BMP-4 to modulate its activity during vertebrate embryonic patterning. Development 125:857–867

Neave B, Holder N, Patient R (1997) A graded response to BMP-4 spatially coordinates patterning of the mesoderm and ectoderm in the zebrafish. Mech Dev 62:103–246

Nellen D, Burke R, Struhl G, Basler K (1996) Direct and long-range action of a dpp morphogen gradient. Cell 85:357–368

Nguyen VH, Schmid B, Trout J, Connors SA, Ekker M, Mullins MC (1998) Ventral and lateral regions of the zebrafish gastrula, including the neural crest progenitors, are established by a bmp2b/swirl pathway of genes. Dev Biol 199:93–110

Niehrs C (1996) Mad connection to the nucleus. Nature 381:561–562

Nieuwkoop PD (1969) The formation of mesoderm in Urodelean Amphibians. I Induction by the Endoderm. Roux Arch 162:341–373

Nieuwkoop PD (1973) The organisation centre of the amphibian embryo, its origin, spatial organisation and morphogenetic action. Adv Morph 10:1–39

Nishimatsu S, Suzuki A, Shoda A, Murakami K, Ueno N (1992) Genes for bone morphogenetic proteins are differentially transcribed in early amphibian embryos. Biochem Biophys Res Commun 186:1487–95

Nishimatsu S-I, Thomsen GH (1998) Ventral mesoderm induction and patterning by bone morphogenetic proteins. Mech Dev 74:75–88

Onichtchouk D, Gawantka V, Dosch R, Delius H, Hirschfeld K, Blumenstock C, Niehrs C (1996) The Xvent-2 homeobox gene is part of the BMP-4 signaling pathway controlling dorsoventral patterning of *Xenopus* mesoderm. Development 122:3045–3053

Onichtchouk D, Glinka A, Niehrs C (1998) Requirement for Xvent-1 and Xvent-2 gene function in dorsoventral patterning of *Xenopus* mesoderm. Development 125:1447–1456

Papalopulu N, Kintner C (1996) A *Xenopus* gene, Xbr-1, defines a novel class of homeobox genes and is expressed in the dorsal ciliary margin of the eye. Dev Biol 174:104–114

Piccolo S, Sasai Y, Lu B, De Robertis EM (1996) Dorsoventral patterning in *Xenopus*: inhibition of ventral signals by direct binding of chordin to BMP-4. Cell 86:589–598

Piccolo S, Agius E, Lu B, Goodman S, Dale L, De Robertis EM (1997) Cleavage of chordin by Xolloid metalloprotease suggests a role for proteolytic processing in the regulation of Spemann organizer activity. Cell 91:407–416

ReemKalma Y, Lamb T, Frank D (1995) Competition between noggin and BMP-4 activities may regulate dorsalization during *Xenopus* development. Proc Natl Acad Sci USA 92:12141–12145

Reilly KM, Melton DA (1996) Short-range signaling by candidate morphogens of the TGF-beta family and evidence for a relay mechanism of induction. Cell 86:743–754

Salic AN, Kroll KL, Evans LM, Kirschner MW (1997) Sizzled: a secreted Xwnt8 antagonist expressed in the ventral marginal zone of *Xenopus* embryos. Development 124:4739–4748

Sasai Y, Lu B, Steinbeisser H, Geissert D, Gont LK, De Robertis EM (1994) *Xenopus* chordin: a novel dorsalizing factor activated by organizer-specific homeobox genes. Cell 79:779–90

Sasai Y, Lu B, Steinbeisser H, De Robertis EM (1995) Regulation of neural induction by the Chd and Bmp-4 antagonistic patterning signals in *Xenopus*. Nature 376:333–6

Sasai Y, Lu B, Piccolo S, De Robertis EM (1996) Endoderm induction by the organizer-secreted factors chordin and noggin in *Xenopus* animal caps. EMBO J. 15:4547–4555

Sauman I, Berry SJ (1994) An actin infrastructure is associated with eukaryotic chromosomes: structural and functional significance. Eur J Cell Biol 64:348–56

Schmidt J, Francois V, Bier E, Kimelman D (1995a) *Drosophila* short gastrulation induces an ectopic axis in *Xenopus*: evidence for conserved mechanisms of dorsal-ventral patterning. Development 121:4319–4328

Schmidt JE, Suzuki A, Ueno N, Kimelman D (1995b) Localized BMP-4 mediates dorsal/ventral patterning in the early *Xenopus* embryo. Dev Biol 169:37–50

Schmidt JE, von Dassow G, Kimelman D (1996) Regulation of dorsal-ventral patterning: the ventralizing effects of the novel *Xenopus* homeobox gene Vox. Development 122:1711–1721

Sekelsky JJ, Newfeld SJ, Raftery LA, Chartoff EH, Gelbart WM (1995) Genetic characterization and cloning of Mothers against dpp, a gene required for decapentaplegic function in *Drosophila melanogaster*. Genetics 139:1347–1358

Shih J, Fraser SE (1996) Characterizing the zebrafish organizer: microsurgical analysis at the early-shield stage. Development 122:1313–22

Sive HL (1993) The frog prince-ss: a molecular formula for dorsoventral patterning in *Xenopus*. Genes Dev 7:1–12

Slack JM (1993) Embryonic induction. Mech Dev 41:91–107

Smith WC, Harland RM (1992) Expression cloning of noggin, a new dorsalizing factor localized to the Spemann organizer in *Xenopus* embryos. Cell 70:829–40

Smith WC, Knecht AK, Wu M, Harland RM (1993) Secreted noggin protein mimics the Spemann organizer in dorsalizing *Xenopus* mesoderm. Nature 361:547–9

Staehling-Hampton K, Laughon AS, Hoffmann FM (1995) A *Drosophila* protein related to the human zinc finger transcription factor PRDII/MBPI/HIV-EP1 is required for dpp signaling. Development 121:3393–3403

Steinbeisser H, Fainsod A, Niehrs C, Sasai Y, De Robertis EM (1995) The role of gsc and BMP-4 in dorsal-ventral patterning of the marginal zone in *Xenopus*: a loss-of-function study using antisense RNA. EMBO J 14:5230–43

Storey KG, Crossley JM, De RE, Norris WE, Stern CD (1992) Neural induction and regionalisation in the chick embryo. Development 114:729–41

Sturtevant MA, Biehs B, Marin E, Bier E (1997) The spalt gene links the A/P compartment boundary to a linear adult structure in the *Drosophila* wing. Development 124:21–32

Suzuki A, Thies RS, Yamaji N, Song JJ, Wozney JM, Murakami K, Ueno N (1994) A truncated bone morphogenetic protein receptor affects dorsal-ventral patterning in the early *Xenopus* embryo. Proc Natl Acad Sci USA 91:10255–9

Suzuki A, Ueno N, Hemmati-Brivanlou A (1997) *Xenopus* msx1 mediates epidermal induction and neural inhibition by BMP4. Development 124:3037–44

Suzuki A, Chang C, Yingling JM, Wang XF, Hemmati BA (1997a) Smad5 induces ventral fates in *Xenopus* embryo. Dev Biol 184:402–5

Suzuki A, Kaneko E, Maeda J, Ueno N (1997b) Mesoderm induction by BMP-4 and –7 heterodimers. Biochem Biophys Res Commun 232:153–6

Suzuki A, Kaneko E, Ueno N, Hemmati-Brivanlou A (1997c) Regulation of epidermal induction by BMP2 and BMP7 signaling. Dev Biol 189:112–22

Takase M, Imamura T, Sampath TK, Takeda K, Ichijo H, Miyazono K, Kawabata M (1998) Induction of Smad6 mRNA by bone morphogenetic proteins. Biochem Biophys Res Commun 244:26–9

Thomsen GH (1996) *Xenopus* mothers against decapentaplegic is an embryonic ventralizing agent that acts downstream of the BMP2/4 receptor. Development 122:2359–2366

Tidman-Ault C, Dirksen ML, Jamrich M (1996) A novel homeobox gene PV.1 mediates induction of ventral mesoderm in *Xenopus* embryos. Proc Natl Acad Sci USA 93:6415–6420

Wang S, Krinks M, Kleinwaks L, Moos MJ (1997) A novel *Xenopus* homologue of bone morphogenetic protein-7 (BMP-7). Genes Funct 1:259–71

Wharton KA, Ray RP, Gelbart WM (1993) An activity gradient of decapentaplegic is necessary for the specification of dorsal pattern elements in the *Drosophila* embryo. Development 117:807–822

Wilson D, Sheng G, Lecuit T, Dostatni N, Desplan C (1993) Cooperative dimerization of paired class homeo domains on DNA. Genes Dev 7:2120–2134

Wilson PA, Hemmati-Brivanlou A (1995) Induction of epidermis and inhibition of neural fate by Bmp-4. Nature 376:331–3

Wilson PA, Lagna G, Suzuki A, Hemmati-Brivanlou A (1997) Concentration-dependent patterning of the *Xenopus* ectoderm by BMP4 and its signal transducer Smad1. Development 124:3177–84

Winnier G, Blessing M, Labosky PA, Hogan BLM (1995) Bone morphogenetic protein 4 is required for mesoderm formation and patterning in the mouse. Genes Dev 9:2105–2116

Zhang H, Bradley A (1996) Mice deficient for BMP2 are nonviable and have defects in amnion/chorion and cardiac development. Development 122:2977–86

Zhang Y, Musci T, Derynck R (1997) The tumor suppressor Smad4/DPC 4 as a central mediator of Smad function. Curr Biol 7:270–6

Zhu Y, Richardson JA, Parada LF, Graff JM (1998) Smad3 mutant mice develop metastatic colorectal cancer. Cell 94:703–714

Zimmerman LB, De Jesús-Escobar J-E, Harland RM (1996) The Spemann organizer signal noggin binds and inactivates bone morphogenetic protein-4. Cell 86:599–606

11 The Indian Hedgehog – PTHrP System in Bone Development

A. Vortkamp

11.1 Introduction

During embryonic development the bones of the vertebrate skeleton develop by two different mechanisms. Most of the bones of the skull are formed by intramembranous ossification in which mesenchymal cells directly differentiate into osteoblasts. The second mechanism, called endochondral ossification, is used to form the bones of the axial and appendicular skeleton as well as some of the facial bones. Endochondral ossification (Erlebacher et al. 1995; Hinchcliffe and Johnson 1980) starts with mesenchymal cells that aggregate and differentiate into chondrocytes, thus forming cartilage elements which serve as templates for the later bones. Cells in the middle of these cartilage elements start to differentiate into hypertrophic chondrocytes, a step that is necessary for the invasion of blood vessels and the subsequent replacement of cartilage by bone. The chondrocytes in the cartilage elements are surrounded by a thin layer of flattened cells, the perichondrium. In parallel with the differentiation of hypertrophic chondrocytes, the perichondrium flanking the differentiating and hypertrophic chondrocytes differentiates into an osteoblast-containing periosteum, which secretes a layer of primary bone, the bone collar. Signals from the perichondrium/periosteum are thought to interact with signals from the cartilage itself in correlating hypertrophic differentiation with the differentiation of the perichondrium into a periosteum. To understand the process of endochondral ossification it is necessary to identify the specific signals regulating the different steps in this process and to analyze their interactions.

11.2 Indian Hedgehog Signaling During Chondrocyte Differentiation

11.2.1 Expression of Ihh

Recently, we and others (Bitgood and McMahon 1995; Vortkamp et al. 1996) have identified a secreted signaling factor, Indian Hedgehog (Ihh), which is expressed in the developing cartilage elements. Comparative expression analysis with chondrocyte markers (Col-II marks all chondrocytes, whereas Col-X specifically marks hypertrophic chondrocytes) reveals that Ihh expression starts during the early steps of bone development in mice and chick. At the early stages Ihh is expressed in the center of the condensing cartilage elements shortly after the chondrocytes are formed. With the onset of hypertrophic differentiation, Ihh expression becomes restricted to the differentiating chondrocytes. At this stage, proliferating and terminally differentiated, hypertrophic chondrocytes do not express Ihh. Therefore, Ihh can be regarded as being specifically expressed in chondrocytes that are actively undergoing hypertrophic differentiation.

11.2.2 Expression of Ptc and Gli

In a variety of organisms, hedgehog molecules have been shown to act through a conserved signaling pathway including homologues of Patched (Ptc) and Gli. Ptc has been identified as a subunit of the hedgehog receptor complex (Goodrich et al. 1996; Marigo et al. 1996a; Marigo et al. 1996c; Stone et al. 1996), and Gli (Marigo et al. 1996b; Ruppert et al. 1988), is a member of a zinc finger transcription factor family. Both genes have been shown to be upregulated in target tissues responding to hedgehog signaling (reviewed in Tabin and McMahon 1997). During the early steps of endochondral ossification, both Ptc and Gli are expressed in the condensing chondrocytes. Their expression domains overlap with, but are wider than that of Ihh. (Vortkamp et al. 1998). With the refinement of the cartilage elements and the subsequent onset of hypertrophic differentiation, Ptc and Gli become strongly expressed in the developing perichondrium flanking the Ihh expression domain. However, weaker expression can still be found in the proliferating chondrocytes distal to the Ihh expression domain (Vortkamp et al. 1996; Vortkamp et al. 1998). These expression patterns suggest that Ihh signals in two directions: to the flanking perichondrium and to the proliferating and less differentiated chondrocytes.

11.3 Ihh and PTHrP Interact in a Negative Feedback Loop Regulating Chondrocyte Differentiation

11.3.1 Ihh Misexpression

To analyze the role of Ihh during endochondral ossification, we used a retroviral vector system to misexpress Ihh in the limbs of developing chick embryos. The limbs were infected at the onset of limb outgrowth at stage Hamburger-Hamilton 22 (HH 22) (Hamburger and Hamilton 1951), when the cartilage condensations start to form. The effect of Ihh misexpression was analyzed at stages of hypertrophic differentiation, at stage HH 34 and stage HH 36. These experiments revealed that misexpression of Ihh disrupts the process of endochondral ossification by blocking hypertrophic differentiation of chondrocytes. Instead, the chondrocytes remain in a proliferating state. More detailed analysis showed that Ihh acts in a negative feedback loop preventing not only the differentiation of hypertrophic chondrocytes but also the differentiation of the Ihh expressing prehypertrophic chondrocytes themselves (Vortkamp et al. 1996).

11.3.2 Interaction of Ihh and PTHrP

The negative effect of Ihh on chondrocyte differentiation seems to be mediated by another secreted factor from the perichondrium, Parathyroid Hormone related Protein (PTHrP), (Amizuka et al. 1994; Karaplis et al. 1994; Lee et al. 1995; Lee et al. 1996; Suva et al. 1987; Weir et al. 1996). PTHrP is most strongly expressed in the joint region of the developing cartilage elements, whereas its receptor, the PTH/PTHrP receptor (Juppner et al. 1991; Lanske et al. 1996), is expressed overlapping with and further distal to the domain of Ihh expression. Retroviral misexpression of Ihh during chick development demonstrated that PTHrP expression is upregulated by Ihh, indicating that PTHrP acts downstream of Ihh in regulating chondrocyte differentiation (Vortkamp et al. 1996). To further investigate the interaction of these signaling molecules, a mouse limb culture system was used to compare the effect of hedgehog protein on chondrocyte differentiation in limbs of wildtype mice to that of PTHrP -/- (Vortkamp et al. 1996) or PTH/PTHrP receptor -/- mice (Lanske et al. 1996). As in the chick, hedgehog protein blocked chondrocyte differentiation in wild-type mouse limbs, whereas no block of hypertrophic differentiation could be detected in limbs of PTHrP -/- or PTH/PTHrP receptor -/- limbs. These experiments strongly suggest that PTHrP signaling is not only regulated by Ihh but is also necessary to mediate the effect of Ihh on chondrocyte differentiation (Lanske et al. 1996; Vortkamp et al. 1996).

11.3.3 Ihh and PTHrP
Regulate the Pace of Chondrocyte Differentiation

What is the role of Ihh and PTHrP signaling during chondrocyte differentiation? One way to interpret the results described above is a model in which Ihh and PTHrP interact in a negative feedback loop regulating the

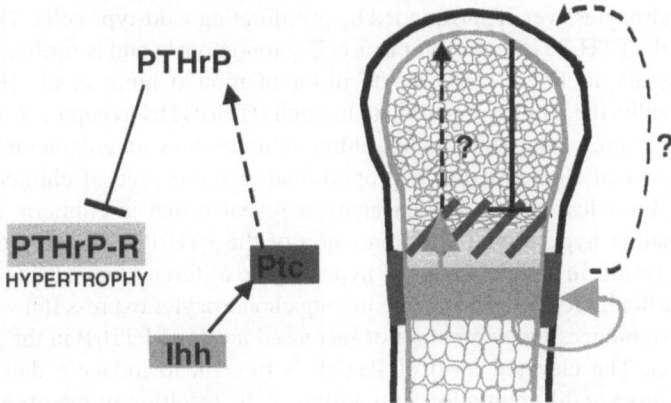

Fig. 1. Ihh and PTHrP form a negative feedback loop regulating chondrocyte differentiation. Ihh (*gray*) expressed from the prehypertrophic chondrocytes signals to the perichondrium and to the proliferating chondrocytes as indicated by the expression of the receptor Ptc (*dark gray*). Ihh expression results in the upregulation of PTHrP (*very light gray*) in the periarticular perichondrium as indicated by the *dashed arrows*. The Ihh signal is mediated either by the perichondrium or by the chondrocytes directly. PTHrP signals back to the PTH/PTHrP receptor (*light gray*) in the proliferating and prehypertrophic chondrocytes to delay chondrocyte differentiation

pace of chondrocyte differentiation (Fig. 1). In this model, the amount of Ihh expressed by the prehypertrophic chondrocytes, would reflect the number of cells undergoing hypertrophic differentiation at a given time. A certain level of Ihh signaling would then induce PTHrP expression in the periarticular perichondrium which in turn would signal back to the PTH/PTHrP receptor, thus preventing the differentiation of additional chondrocytes. Once the differentiating, prehypertrophic chondrocytes are finally differentiated into hypertrophic chondrocytes, they turn off the expression of Ihh. The reduced amount of Ihh signaling would then attenuate the negative feedback loop and allow additional chondrocytes to undergo the differentiation process. This model has recently been supported by elegant experiments analyzing chimeric mice that carry different amounts of PTH/PTHrP receptor -/- cells in a wild-type background (Chung et al. 1998). Firstly, these experiments showed that PTH/PTHrP receptor-/- cells differentiate prematurely into hypertrophic chondrocytes even if surrounded by proliferating wild-type cells. There-fore the PTH/PTHrP receptor acts cell autonomously and is furthermore necessary to block chondrocyte differentiation (Chung et al. 1998). Secondly, the authors found that although PTH/PTHrP receptor -/- mice have severely shortened bones, chimeric mice show an enlargement of the skeletal elements that is proportional to the degree of chimerism. Morphological and molecular analysis revealed that in chimeric mice premature hypertrophic differentiation of the PTH/PTHrP receptor -/-cells results in a second zone of hypertrophic differentiation close to the periarticular region. The differentiating chondrocytes express Ihh which in turn induces the expression of increased levels of PTHrP in the joint region. The elevated level of PTHrP is thought to induce a delay of chondrocyte differentiation in wild-type cells resulting in the observed enlargement of the skeletal elements.

Summarizing, experiments both in chick and mice demonstrated that Ihh and PTHrP interact to regulate chondrocyte differentiation. How-ever, whether the interaction of Ihh and PTHrP is direct or indirect, and how the Ihh signal is transmitted to the joint region to induce PTHrP has still to be elucidated. Ptc as a receptor for Ihh seems to have an impor-tant function in this process and it is necessary to analyze the role of the two Ptc expression domains in regulating chondrocyte differentiation in more detail to fully understand Ihh signaling.

11.4 The Ihh/PTHrP Negative Feedback Loop Regulates Postnatal Bone Growth and Fracture Repair

11.4.1 Postnatal Bone Growth

Endochondral ossification is not only an embryonic process, but contin-ues until the proliferating chondrocytes are completely replaced by bone and only articular cartilage remains at either end of the skeletal ele-ments. Mice and other mammals differ slightly as they develop secon-dary ossification centers at the distal ends of the skeletal elements, leaving a zone of proliferating and differentiating chondrocytes, the epiphysial growth plate, in between the two ossification areas (Floyd et al. 1987). The growth plate is responsible for postnatal bone growth

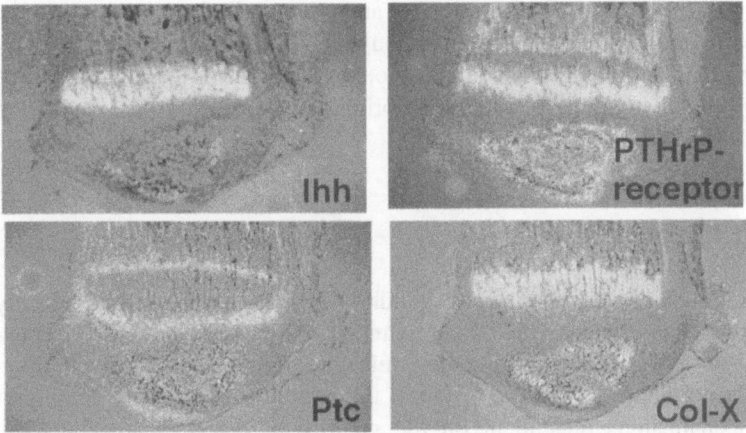

Fig. 2. The genes of the Ihh/PTHrP feedback loop are expressed in the postnatal growth plate. In situ hybridization shows the expression of Ihh, PTH/PTHrP receptor, Ptc, and Col-X in the primary and secondary ossification center in the growth plate of a day 13 mouse tibia

until the ossification centers connect during puberty, resulting in a bony skeletal element which is again flanked by caps of articulate cartilage.

We analyzed the expression of Ihh and its interactors during postnatal bone growth in mice (Fig. 2) and found that Ihh and the PTH/PTHrP receptor are still expressed in the differentiating, prehypertrophic chondrocytes at least up to postnatal day 21 (P21) (Iwasaki et al. 1997; Vortkamp et al. 1998). Ptc and Gli are strongly expressed in the resting and proliferating chondrocytes distal to the Ihh expressing cells in the metaphysis. In addition we found a second domain of strong Ptc and Gli expression in the zone of active ossification, directly flanking the hypertrophic chondrocytes (Vortkamp et al. 1998). The expression of Ptc and Gli in both the proliferating chondrocytes and the zone of ossification implies that hedgehog signaling regulates both chondrocyte differentiation and ossification. In this respect, the early expression of Ptc and Gli in the perichondrium/periosteum could be involved in regulating the differentiation of the perichondrium into a bone-producing periosteum, whereas the expression in the proliferating chondrocytes might be responsible for regulating chondrocyte differentiation.

Ihh, PTH/PTHrP receptor, Ptc, and Gli are all also expressed in the secondary ossification center, although in very restricted regions, implying that the Ihh/PTHrP feedback loop regulates the differentiation of this postnatal structure as well as the primary ossification process (Pathi et al. 1999).

11.4.2 Fracture Repair

A second process where endochondral ossification can take place during adult life is the process of fracture repair. Whereas there are two mechanisms to form bone during embryonic development, fracture repair takes place independent of the origin of the injured bone. Instead, the repair mechanism is dependent on the severity of the fracture. Stabilized fractures without gaps heal by direct differentiation of bone from mesenchymal cells with no, or relatively small, amounts of cartilage formation, whereas unstabilized fractures or fractures with gaps form large cartilaginous calluses which are then replaced by bone. It has been shown previously that the morphological sequence of chondrocyte differentiation, as well as the expression of extracellular markers during fracture repair, resembles the different steps of embryonic endochondral ossification, although the tissues in the various differentiation stages seem to be less organized (Bolander 1992; Sandberg et al. 1993).

In addition to several cartilage markers, the chondrocytes of the wound-healing callus express Ihh and PTH/PTHrP receptor in those chondrocytes undergoing hypertrophic differentiation (Fig. 3). As in the postnatal growth plate, Gli and Ptc are expressed in the undifferentiated chondrocytes at the border of the cartilage callus as well as in the zone of active ossification (Vortkamp et al. 1998).

Summarizing these expression data we can conclude that the same mechanisms that regulate embryonic endochondral ossification are used to control postnatal bone growth and can be reactivated during fracture repair.

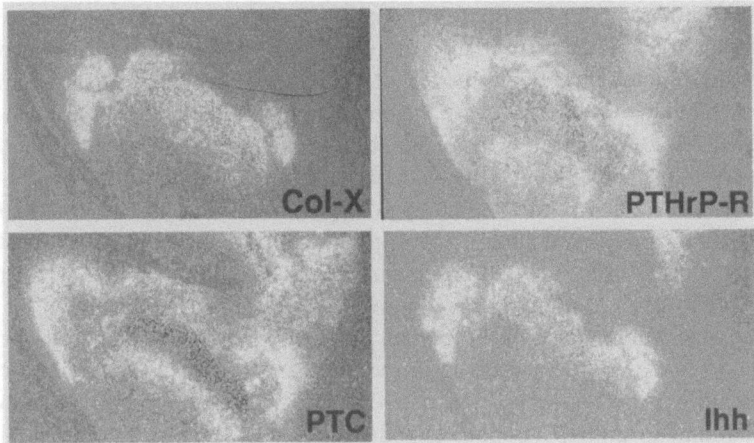

Fig. 3. The genes of the Ihh/PTHrP feedback loop are expressed during fracture repair. In situ hybridization shows the expression of Ihh, PTH/PTHrP receptor, Ptc, and Col-X in the cartilage callus on day 7 after the induction of the fracture

11.5 Interaction of Ihh and BMP Signaling

11.5.1 BMP Expression During Chondrocyte Differentiation

To gain deeper insight into the process of endochondral ossification one has to appreciate how the Ihh/PTHrP signaling pathway interacts with other signals regulating bone development. Bone morphogenetic proteins (BMPs), a subgroup of the conserved TGF-β superfamily of secreted proteins (Hogan 1996; Kingsley 1994a; Kingsley 1994b), are another group of signaling molecules that have been implicated in regulating bone formation. During the early stages of bone development, BMP genes and their receptors are expressed in distinct cell types within the cartilage elements. For example, BMP2, BMP4, BMP5, and BMP7 are expressed in the perichondrium surrounding the skeletal elements (Jones et al. 1991; Kingsley 1994b; Lyons et al. 1990; Zou et al. 1997), whereas BMP6 and the BMP receptor IA (BMPR-IA) are expressed in the hypertrophic and prehypertrophic chondrocytes (Kawakami et al. 1996; Kingsley 1994a; Vortkamp et al. 1996).

Comparative analysis of Ihh and BMP expression at chondrocyte differentiation stages in the developing chick embryo (day 8) revealed changes in the level of expression of BMP2, BMP4, BMP5, BMP6, and BMP7 that correlate well with the borders of the Ihh expression domain. BMP7 is expressed in the perichondrium flanking the hypertrophic and prehypertrophic chondrocytes, but not distal to the border of Ihh expression. BMP2 and BMP4 are strongly expressed in the same region, and weaker expression domains reach distal towards the ends of the skeletal elements. BMP5 shows complementary expression levels with a weaker expression level flanking the Ihh expression domain and a higher expression level outside the borders of the Ihh expression domain (Pathi et al. 1999). BMP6 is expressed in the prehypertrophic and hypertrophic chondrocytes (Vortkamp et al. 1996) whereas the BMPR-IA expression is found in prehypertrophic chondrocytes only (Zou et al. 1997). In addition, at least BMP2, BMP4, and BMPR-IA are strongly expressed in the perichondrium of the joint region (Pathi et al. 1999; Zou et al. 1997). The correlation of BMP expression with that of Ihh suggests that the two signaling pathways might interact with each other in regulating cartilage development.

11.5.2 Ihh Regulates BMP Expression

It has been shown that misexpression of BMP4 (Duprez et al. 1996), as well as misexpression of a constitutively active form of BMPR-IA (Zou et al. 1997), in wings of developing chick embryos results in a massive overgrowth of the wing skeletal elements. The morphological and molecular analysis of skeletal elements infected with a constitutively active BMPR-IA revealed that activation of BMP signaling prevents chondrocyte differentiation in a manner similar to that of misexpression of Ihh (Zou et al. 1997). Like Ihh misexpression, ecotopic BMP signaling activates PTHrP expression in the periarticular perichondrium implicating an interaction of the Ihh and BMP pathways. The upregulation of PTHrP in the perichondrium is not accompanied by an upregulation of Ihh, indicating that BMP signaling acts downstream of Ihh (Zou et al. 1997).

To test this hypothesis we analyzed the effect of Ihh misexpression on the expression of various BMP genes in the perichondrium. In principal there are two ways BMP expression can respond to Ihh misexpression. If Ihh directly regulates the expression of BMP genes, ectopic Ihh expression should lead to a BMP expression pattern throughout the perichondrium that reflects the pattern normally flanking the Ihh expression domain in the prehypertrophic chondrocytes. If only dependent on Ihh, this expression pattern should be independent of the stage of chondrocyte differentiation. Alternatively, the correlation between BMP and Ihh expression domains could be established by other signals from the chondrocytes, which regulate Ihh and BMP expression independently. In that case, ectopic expression of Ihh should lead to a BMP expression level throughout the perichondrium which reflects the differentiation state of the adjacent chondrocytes. As Ihh misexpression blocks chondrocyte differentiation before they reach the prehypertrophic state, the expression of the BMP genes should resemble the expression that is normally found flanking the proliferating cells.

We found by misexpression of Ihh that BMP5, BMP6, and BMP7 are not directly regulated by Ihh, but rather their expression seems to be dependent on the differentiation state of the chondrocytes: BMP6 and BMP7 expression is downregulated and BMP5 expression is upregulated.

In contrast, the expression of BMP2 and BMP4 is strongly upregulated upon Ihh misexpression independent of the state of chondrocyte differntiation. These BMP genes are normally most strongly expressed in the perichondrium flanking the Ihh expressing prehypertrophic and hypertrophic chondrocytes, and weaker outside this region. Ihh misexpression results in strong induction of BMP2 and BMP4 expression in the perichondrium flanking proliferating chondrocytes, when these express high levels of Ihh. Therefore, these two BMP genes are good candidates for being direct targets of Ihh signaling.

Ihh expression has been shown to regulate PTHrP expression in the periarticular perichondrium (Vortkamp et al. 1996). However, as hedgehog molecules are not thought to travel over long distances, secondary molecules seem to be required to mediate the Ihh signal from the prehypertrophic chondrocytes to the periarticular perichondrium. In combination with the results of Zou et al. (1997), BMP2 and BMP4 but not BMP5, BMP6, or BMP7 might be good candidates to serve as these secondary signals. They could then act through BMPR-IA, which is expressed in the perichondrium of the joint, to induce PTHrP (Fig. 4).

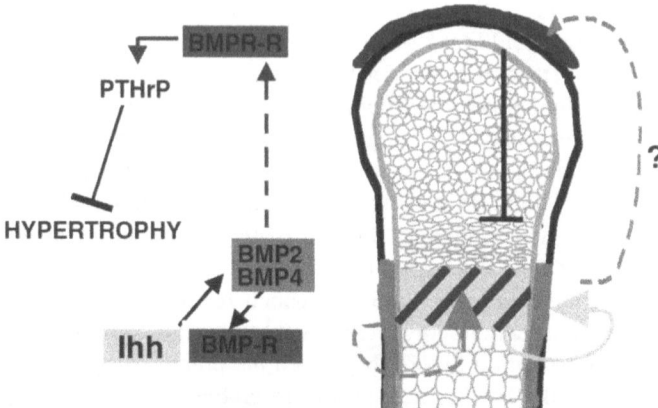

Fig. 4. Interaction of Ihh/PTHrP signaling with BMP signaling. Ihh *light gray)* expression induces the expression of BMP2 and BMP4 (*gray*) in the perichondrium flanking the Ihh expression domain. BMP signaling could then be involved in mediating the Ihh signal to the periarticular perichondrium to induce PTHrP (*very light gray*) via the BMP-R (*dark gray*) expressed in the joint region. Alternatively, or in addition, BMPs might signal back to the BMP receptor in the prehypertrophic chondrocytes to correlate hypertrophic differentiation of chondrocytes with the differentiation of the perichondrium into a periosteum

Alternatively, or additionally, BMP signaling might be involved in regulating the differentiation of the perichondrium into a periosteum. In parallel with the differentiation of hypertrophic chondrocytes, the perichondrium differentiates into a periosteum, containing bone-producing osteoblasts. The border of the periosteum correlates well with the border of Ihh expression in the chondrocytes suggesting a correlation between the two processes. Interestingly, the periosteum expresses elevated levels of BMP2 and BMP4 as well as BMP7. These could act through BMPR-IA, which is strongly expressed in the adjacent prehypertrophic chondrocytes. Therefore upregulation of BMP expression in the periosteum induced by Ihh from the prehypertrophic chondrocytes could then signal back to the prehypertrophic cells to coordinate hypertrophic differentiation with the differentiation of the periosteum. It would be interesting to analyze whether BMP signaling from the periosteum can reciprocally influence the expression of Ihh in the differentiation chondrocytes.

11.5.3 BMP Antagonists in Developing Skeletal Elements

BMP signaling can be regulated by several antagonists including Noggin and Chordin (Hirsinger et al. 1997; Marcelle et al. 1997; Reshef et al. 1998; Sasai et al. 1994; Smith and Harland, 1992; Smith et al. 1993). Both are secreted molecules which can bind to different BMPs and prevent them from binding to their receptors (Piccolo et al. 1997; Piccolo et al. 1996; Zimmerman et al. 1996). Recently it has been shown that Noggin and Chordin are both expressed in the developing cartilage elements. Noggin is initially expressed in the early cartilage condensations. As the skeletal elements develop, Noggin expression becomes progressively restricted to the ends of the cartilage elements distal to the proliferating chondrocytes. In addition, low levels of Noggin expression can be found in the proliferating and differentiating chondrocytes, whereas the hypertrophic chondrocytes no longer express Noggin (Capdevila and Johnson 1998; McMahon et al. 1998; Merino et al. 1998; Pathi et al. 1999). At cartilage differentiation stages, Chordin is expressed in the periarticular region of the cartilage elements in a domain overlapping with, but wider than that expressing PTHrP (Pathi et al. 1999). The similarity of the expression domains of PTHrP and Chordin indicates that Chordin might be regulated by Ihh in a similar manner as was shown for PTHrP. We analyzed the expression of Chordin after Ihh misexpression in embryonic chick limbs on E8 and found that, like PTHrP, chordin was strongly upregulated in the periarticular perichondrium. We also analyzed the expression of Noggin after Ihh misexpression and found Noggin throughout the cartilage element, most likely as a result of the block of chondrocyte differentiation (Pathi et al. 1999).

Together these data indicate that Ihh regulates BMP signaling, not only by regulating the expression of several BMP genes but also by influencing the expression of antagonists of BMP signaling at least Noggin and Chordin.

11.5.4 Noggin Misexpression

To further explore the interaction of Ihh/PTHrP signaling with that of
BMPs we have started to use Noggin misexpression as a tool to block
BMP signaling during chondrocyte differentiation. We found that BMP
signaling seems to act at least at three different points during cartilage
differentiation. Firstly, blocking of BMP signaling before limb outgrowth can completely block the formation of the cartilage elements,
indicating the importance of BMP signals for inducing endochondral
ossification (Capdevila and Johnson 1998; McMahon et al. 1998; Merino et al. 1998; Pathi et al. 1999).

Secondly, misexpression of Noggin after the onset of limb outgrowth
results in extremely shortened cartilage elements. The infection protocol
chosen was such that a considerable expression of Noggin should be
restricted to time points after the condensations have been formed. It has
been hypothesized that BMPs might be involved in mediating the effect
of Ihh signaling in blocking chondrocyte differentiation (Pathi et al.
1999; Zou et al. 1997). If this is true, blocking BMP signaling should
result in advanced hypertrophic differentiation. Instead we found that
the block of BMP signaling at these stages seems to inhibit both chondrocyte proliferation and differentiation. This result can, however, be a
consequence of the third effect we observed after Noggin misexpression: Noggin strongly upregulates BMP4 expression in the perichondrium of the infected skeletal elements especially in the joint region.
Therefore, BMP signaling seems to be regulated by a strong autoregulatory component where blocking of BMP signaling upregulates at least
one BMP gene. This negative feedback loop could also antagonize the
effect of Noggin misexpression on chondrocyte differentiation.

11.6 Conclusions

Endochondral ossification is a complex multistep process which is
regulated by different control systems. Ihh and PTHrP are two signaling
molecules that act together in a negative feedback loop to regulate the
pace of chondrocyte differentiation, one of the critical steps of endochondral ossification. Although both secreted factors were identified as
regulators of the embryonic process of bone development, both continue

to be expressed after birth and regulate postnatal bone growth as well as fracture repair. This demonstrates that the analysis of embryonic signaling systems will not only result in an understanding of embryonic bone development, but will also have impact on possible forms of treatment for postnatal bone diseases and bone fractures.

It is expected that the Ihh/PTHrP pathway is only part of a complex signaling system. Intensive studies are necessary to identify the other players of this signaling network, to analyze their function and to study their interactions. BMPs are one group of factors that have been anticipated to play a crucial role in regulating bone formation and several lines of evidence indicate that in addition to inducing the onset of bone formation, BMPs might act at several stages during chondrocyte differentiation. Activation of BMP signaling has been shown to result in similar phenotypes as those seen after overexpression of Ihh. At least two BMP genes, BMP2 and BMP4, are good candidates to be directly regulated by Ihh. In addition, expression analyses of two BMP antagonists, Noggin and Chordin, indicate that BMP signaling during bone development might interact with that of the Ihh/PTHrP system. However, the exact interaction of the two signaling pathways requires further analysis.

References

Amizuka N, Warshawsky H, Henderson JE, Goltzman D, Karaplis AC (1994) Parathyroid hormone-related peptide-depleted mice show abnormal epiphyseal cartilage development and altered endochondral bone formation. J Cell Biol 126:1611–1623

Bitgood MJ, McMahon AP (1995) Hedgehog and Bmp genes are coexpressed at many diverse sites of cell–cell interaction in the mouse embryo. Dev Biol 172:126–138

Bolander ME (1992) Regulation of fracture repair by growth factors. Proc Soc Exp Biol Med 200:165–170

Capdevila J, Johnson RL (1998) Endogenous and ectopic expression of noggin suggests a conserved mechanism for regulation of BMP function during limb and somite patterning. Dev Biol 197:205–217

Chung UI, Lanske B, Lee K, Li E, Kronenberg H (1998) The parathyroid hormone/parathyroid hormone-related peptide receptor coordinates endochondral bone development by directly controlling chondrocyte differentiation. Proc Natl Acad Sci USA 95:13030–13035

Duprez D, Bell EJ, Richardson MK, Archer CW, Wolpert L, Brickell PM, Francis-West PH (1996) Overexpression of BMP-2 and BMP-4 alters the size and shape of developing skeletal elements in the chick limb. Mech Dev 57:145–157

Erlebacher A, Filvaroff EH, Gitelman SE, Derynck R (1995) Toward a molecular understanding of skeletal development. Cell 80:371–378

Floyd WED, Zaleske DJ, Schiller AL, Trahan C, Mankin HJ (1987) Vascular events associated with the appearance of the secondary center of ossification in the murine distal femoral epiphysis. J Bone Joint Surg Am 69:185–190

Goodrich LV, Johnson RL, Milenkovic L, McMahon JA, Scott MP (1996) Conservation of the hedgehog/patched signaling pathway from flies to mice: induction of a mouse patched gene by hedgehog. Genes Dev 10:301–312

Hamburger V, Hamilton HL (1951) A series of normal stages in the development of the chick embryo. J Exp Morph 88:49–92

Hinchcliffe JR, Johnson DR (1980) The development of the Vertebrate limb, Oxford University Press, New York

Hirsinger E, Duprez D, Jouve C, Malapert P, Cooke J, Pourquie O (1997) Noggin acts downstream of Wnt and Sonic Hedgehog to antagonize BMP4 in avian somite patterning. Development 124:4605–4614

Hogan BLM (1996) Bone morphogenetic proteins: multifunctional regulators of vertebrate development. Genes Dev 10:1580–1594

Iwasaki M, Le AX, Helms JA (1997) Expression of Indian hedgehog, bone morphogenetic protein 6 and gli during skeletal morphogenesis. Mech Dev 69:197–202

Jones CM, Lyons KM, Hogan BL (1991) Involvement of Bone Morphogenetic Protein-4 (BMP-4) and Vgr-1 in morphogenesis and neurogenesis in the mouse. Development 111:531–542

Juppner H, Abou-Samra AB, Freeman M, Kong XF, Schipani E, Richards J, Kolakowski L Jr, Hock J, Potts J Jr, Kronenberg HM (1991) A G protein-linked receptor for parathyroid hormone and parathyroid hormone-related peptide. Science 254:1024–1026

Karaplis AC, Luz A, Glowacki J, Bronson RT, Tybulewicz VL, Kronenberg HM, Mulligan RC (1994) Lethal skeletal dysplasia from targeted disruption of the parathyroid hormone-related peptide gene. Genes Dev 8:277–289

Kawakami Y, Ishikawa T, Shimabara M, Tanda N, Enomoto-Iwamoto M, Iwamoto M, Kuwana T, Ueki A, Noji S, Nohno T (1996) BMP signaling during bone pattern determination in the developing limb. Development 122:3557–3566

Kingsley DM (1994a) The TGF-beta superfamily: new members, new receptors, and new genetic tests of function in different organisms. Genes Dev 8:133–146

Kingsley DM (1994b) What do BMPs do in mammals? Clues from the mouse short-ear mutation. Trends Genet 10:16–21

Lanske B, Karaplis AC, Lee K, Luz A, Vortkamp A, Pirro A, Karperien M, Defize LHK, Ho C, Mulligan RC, Abou-Samra AB, Juppner H, Segre GV, Kronenberg HM (1996) PTH/PTHrP receptor in early development and Indian hedgehog-regulated bone growth [see comments]. Science 273:663–666

Lee K, Deeds JD, Segre GV (1995) Expression of parathyroid hormone-related peptide and its receptor messenger ribonucleic acids during fetal development of rats. Endocrinology 136:453–463

Lee K, Lanske B, Karaplis AC, Deeds JD, Kohno H, Nissenson RA, Kronenberg HM, Segre GV (1996) Parathyroid hormone-related peptide delays terminal differentiation of chondrocytes during endochondral bone development. Endocrinology 137:5109–118

Lyons KM, Pelton RW, Hogan BL (1990) Organogenesis and pattern formation in the mouse: RNA distribution patterns suggest a role for bone morphogenetic protein-2A (BMP-2A) Development 109:833–844

Marcelle C, Stark MR, Bronner-Fraser M (1997) Coordinate actions of BMPs, Wnts, Shh and noggin mediate patterning of the dorsal somite. Development 124:3955–3963

Marigo V, Davey RA, Zuo Y, Cunningham JM, Tabin CJ (1996a) Biochemical evidence that patched is the Hedgehog receptor [see comments]. Nature 384:176–179

Marigo V, Johnson RL, Vortkamp A, Tabin CJ (1996b) Sonic hedgehog differentially regulates expression of GLI and GLI3 during limb development. Dev Biol 180:273–283

Marigo V, Scott MP, Johnson RL, Goodrich LV, Tabin CJ (1996c) Conservation on hedgehog signaling: induction of a chicken patched homolog by Sonic hedgehog in the developing limb. Development 122:1225–1233

McMahon JA, Takada S, Zimmerman LB, Fan CM, Harland RM, McMahon AP (1998) Noggin-mediated antagonism of BMP signaling is required for growth and patterning of the neural tube and somite. Genes Dev 12:1438–1452

Merino R, Ganan Y, Macias D, Economides AN, Sampath KT, Hurle JM (1998) Morphogenesis of digits in the avian limb is controlled by FGFs, TGFbetas, and noggin through BMP signaling. Dev Biol 200:35–45

Pathi S, Rutenberg JB, Johnson RL, Vortkamp A (1999) Interaction of Ihh and BMP/Noggin signaling during cartilage differentiation. Dev Biol 209:239–253

Piccolo S, Agius E, Lu B, Goodman S, Dale L, De Robertis EM (1997) Cleavage of Chordin by Xolloid metalloprotease suggests a role for proteolytic processing in the regulation of Spemann organizer activity. Cell 91:407–146

Piccolo S, Sasai Y, Lu B, De Robertis E (1996) Dorsoventral patterning in *Xenopus*: inhibition of ventral signals by direct binding of chordin to BMP-4. Cell 86:589–598

Reshef R, Maroto M, Lassar AB (1998) Regulation of dorsal somitic cell fates: BMPs and Noggin control the timing and pattern of myogenic regulator expression. Genes Dev 12:290–303

Ruppert JM, Kinzler KW, Wong AJ, Bigner SH, Kao F-T, Law ML, Seuanez HN, O'Brian SJ, Vogelstein B (1988) The GLI-Kruppel famiy of human genes. Mol. Cel. Biol. 8:3104–3313

Sandberg MM, Aro HT, Vuorio EI (1993) Gene expression during bone repair. Clin Orthop 298:292–312

Sasai Y, Lu B, Steinbeisser H, Geissert D, Gont LK, De Robertis E. M (1994) *Xenopus* chordin: a novel dorsalizing factor activated by organizer- specific homeobox genes. Cell 79:779–90

Smith WC, Harland RM (1992) Expression cloning of noggin, a new dorsalizing factor localized to the Spemann organizer in *Xenopus* embryos. Cell 70:829–840

Smith WC, Knecht AK, Wu M, Harland RM (1993) Secreted noggin protein mimics the Spemann organizer in dorsalizing *Xenopus* mesoderm. Nature 361:547–59

Stone DM, Hynes M, Armanini M, Swanson TA, Gu Q, Johnson RL, Scott MP, Pennica D, Goddard A, Phillips H, Noll M, Hooper JE, de Sauvage F, Rosenthal A (1996) The tumour-suppressor gene patched encodes a candidate receptor for Sonic hedgehog [see comments]. Nature 384:129–134

Suva LJ, Winslow GA, Wettenhall RE, Hammonds RG, Moseley JM, Diefenbach-Jagger H, Rodda CP, Kemp BE, Rodriguez H, Chen EY (1987) A parathyroid hormone-related protein implicated in malignant hypercalcemia: cloning and expression. Science 237:893–896

Tabin CJ, McMahon AP (1997) Recent advances in Hedgehog signalling. Trends Cell Biol 7:442–446

Vortkamp A, Lee K, Lanske B, Segre GV, Kronenberg HM, Tabin CJ (1996) Regulation of rate of cartilage differentiation by Indian hedgehog and PTH-related protein [see comments]. Science 273:613–622

Vortkamp A, Pathi S, Peretti GM, Caruso EM, Zaleske DJ, Tabin CJ (1998) Recapitulation of signals regulating embryonic bone formation during postnatal growth and in fracture repair. Mech Dev 71:65–76

Weir EC, Philbrick WM, Amling M, Neff LA, Baron R, Broadus AE (1996) Targeted overexpression of parathyroid hormone-related peptide in chondrocytes causes chondrodysplasia and delayed endochondral bone formation. Proc Natl Acad Sci USA 93:10240–10245

Zimmerman LB, De Jesus-Escobar JM, Harland RM (1996) The Spemann organizer signal noggin binds and inactivates bone morphogenetic protein 4. Cell 86:599–606

Zou H, Wieser R, Massague J, Niswander L (1997) Distinct roles of type I bone morphogenetic protein receptors in the formation and differentiation of cartilage. Genes Dev 11:2191–2203

12 Hedgehog Signaling in Animal Development and Human Disease

E. C. Bailey, M. P. Scott, and R. L. Johnson

12.1 Introduction

Higher eukaryotes have developed a number of mechanisms to address
the challenge of progressing from a single-cell zygote to a multicellular
adult organism. Forming complex structures such as a wing, eye, or a
spinal cord requires precise control of cell growth and differentiation.
Generating spatial patterns in embryogenesis requires communication
between adjacent cells, and secreted proteins are one way by which
neighboring cells relay information. The Hedgehog (Hh) family of
secreted proteins is critical for the embryonic development of many
animals. Hh signaling mechanisms thus provide a window into early
development and pattern formation. Genetic screens using *Drosophila*
have revealed a number of components involved in Hh signaling. For
many of them, their vertebrate counterparts have been found and this

pathway is conserved to a remarkable degree. Furthermore, several components involved in Hh signaling are mutated in human tumors and developmental syndromes, highlighting the importance of this pathway in disease. Recent work has focused on understanding how the various components of the Hh pathway receive and transduce the Hh signal to control cell fate and proliferation decisions. The dissection of the Hh pathway has provided valuable insights into the basic mechanisms of development and disease.

12.2 Reception of the Hh Signal

12.2.1 Functional Studies of Ptc and Smo

A number of genetic and biochemical experiments have led to a model of how Hh signaling is received and regulated (Fig. 1). Hh encodes a secreted glycoprotein that signals to neighboring cells (Lee et al. 1992; Mohler and Vani 1992; Tabata et al. 1992; Tashiro et al. 1993). Cells that are competent to receive Hh signals express two critical components, the membrane proteins Patched (Ptc) and Smoothened (Smo). These proteins have opposing effects on Hh-mediated signaling. Smo, a protein with sequence hallmarks of G-protein coupled receptors, is a positive regulator of Hh-induced transcription (Alcedo et al. 1996; van den Heuvel and Ingham 1996). In contrast, Ptc, a protein with 12 potential membrane-spanning domains (Hooper and Scott 1989; Nakano et al. 1989), prevents the induction of Hh target genes. In the absence of Hh, Ptc is proposed to constitutively inhibit transcription (Hidalgo 1991; Ingham et al. 1991), possibly by associating with and inactivating Smo (Stone et al. 1996). Hh is thought to induce signaling by binding to Ptc and thereby relieving the inhibition of Smo (Chen and Struhl 1996; Marigo et al. 1996; Stone et al. 1996). How Ptc affects Smo, in molecular terms, is not well understood. Apparently, Hh signaling does not dissociate Smo from Ptc since the two proteins remain together in the presence of ligand (Stone et al. 1996). While Smo contains an extracellular cysteine-rich domain (CRD) at its N-terminus that could bind a ligand, none has been identified yet. Smo does not bind Hh (Stone et al. 1996) but it may bind a ligand similar to Wnts, since the CRD of Smo is

A   **B**

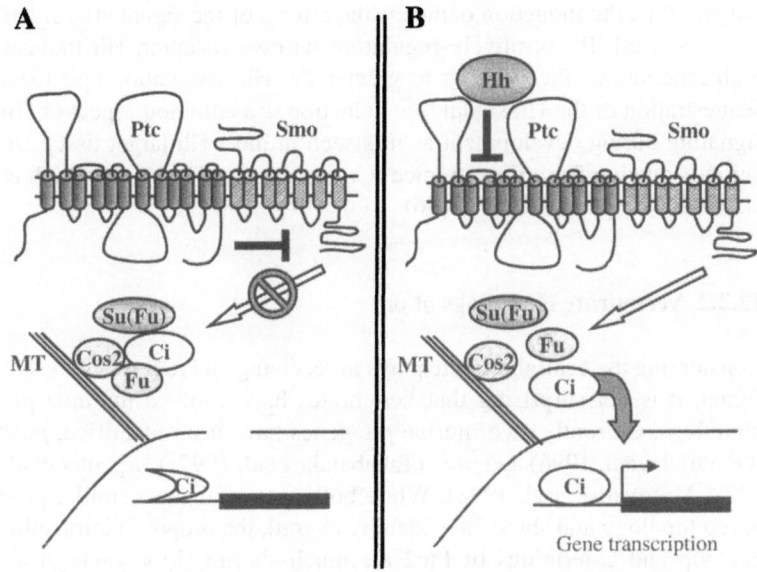

Fig. 1A,B. A current model of Hedgehog signaling. **A** In the absence of Hh ligand, Ptc inhibits Hh-mediated gene transcription by blocking Smo activity. In the cytosol, full length Ci is part of a complex that includes Cos2, Fu, and Su(Fu) that is tethered to microtubules (*MT*). Without signaling, Ci is proteolytically converted into transcriptional repressor that blocks specific gene expression in the nucleus. **B** Hh binding to Ptc relieves the inhibition of Smo and activates signaling. Ci proteolysis is blocked and an activated form of Ci enters the nucleus to induce target gene transcription. It is not known if Ci entry into the nucleus requires the disruption of the complex as depicted

related to that of the Frizzled family of Wnt receptors (Alcedo et al. 1996; van den Heuvel and Ingham 1996).

Ptc regulates Hh signaling in a second way, by binding and sequestering Hh to limit its range of action (Chen and Struhl 1996; Chen and Struhl 1998). When Hh signals to a cell, the amounts of *ptc* transcript and protein are induced from low to high levels (Hidalgo and Ingham 1990; Ingham et al. 1991). Induction of Ptc by Hh has the important consequence of restricting the range of Hh signaling. When cells are unable to raise levels of Ptc in response to Hh, Hh protein (Taylor et al. 1993) and activity (Chen and Struhl 1996) extend beyond their normal range. Thus, Ptc induction dampens the effects of the signal after it has been received. By positively regulating its own receptor, Hh induces high amounts of Ptc, leading to greater Ptc-Hh association and local sequestration of the Hh signal. *ptc* induction is a common aspect of Hh signaling during development as it is seen in most Hh target tissues in animals ranging from flies to mice (Concordet et al. 1996; Goodrich et al. 1996; Marigo and Tabin 1996).

12.2.2 Vertebrate Homologs of *ptc*

Considering the central role Ptc plays in receiving and regulating the Hh signal, it is not surprising that vertebrates have evolved multiple *ptc* homologs. Currently, two murine *ptc* genes have been identified, *ptc1* (Goodrich et al. 1996) and *ptc2* (Takabatake et al. 1997; Carpenter et al. 1998; Motoyama et al. 1998). While both proteins have a similar proposed topology and share 54% identity overall, the proposed intracellular loop and C-terminus of Ptc2 are much shorter (Motoyama et al. 1998). When overexpressed in cultured cells, both proteins bind murine Hh family members and associate with Smo (Marigo et al. 1996; Stone et al. 1996; Carpenter et al. 1998). Apparently, either protein could regulate the signaling of murine Hh family members as both Ptc1 and Ptc2 bind Sonic Hedgehog (Shh), Indian Hedgehog (Ihh), and Desert Hedgehog (Dhh) with nM affinity (Carpenter et al. 1998).

The two *ptc* homologs are expressed in a wide range of tissues during mouse development. *ptc1* is expressed at high levels adjacent to tissues that express Hh family members, suggesting that *ptc1* regulates the signaling of all three Hh members in vivo (Bitgood et al. 1996; Goodrich et al. 1996; Vortkamp et al. 1996). As in *Drosophila*, the induction of *ptc1* suggests that cells have received Hh signals and have raised Ptc1 levels to sequester and restrict the amount of signaling. Genetic studies show that *ptc1* is critical in regulating Shh signaling (Goodrich et al. 1996; Hahn et al. 1998), and other evidence suggests that Ihh and Dhh may likewise mediate their effects through Ptc1 in vivo (Bitgood et al. 1996; Goodrich et al. 1996; Vortkamp et al. 1996). The role of *ptc2* in mouse development is less clear. In general, *ptc2* is expressed in a pattern similar to *ptc1* but at much lower levels (Carpenter et al. 1998;

Motoyama et al. 1998). There are, however, two interesting exceptions, the developing testes and hair follicle.

12.2.3 *ptc1* and *ptc2* May Have Separate Roles in Development

During testis development, *ptc1* and *ptc2* are expressed in distinct cell types, suggesting that these two proteins mediate separate aspects of Dhh signaling. *Dhh* is the only Hh family member expressed in the developing testis and is found in the Sertoli cells (Bitgood et al. 1996), a cell type that supports the developing spermatogonia (Griswold 1995). Targeted deletion of *Dhh* results in normal females but sterile males, indicating that *Dhh* is critical for spermatogenesis (Bitgood et al. 1996). In normal mice, *ptc1* is expressed in the Leydig cells, which secrete androgens and neighbor the Sertoli cells (de Kretser et al. 1998). In the testes of *Dhh* deficient males, *ptc1* expression is no longer detected in the Leydig cells, presumably because Sertoli-derived Dhh is absent. While this suggests Dhh mediates a signal from Sertoli to Leydig cells, the Leydig cells in *Dhh* deficient males appear normal. The purpose of this signaling event and the role of *Ptc1* in testicular development is not yet evident.

In contrast, *ptc2* may have a critical role in spermatogenesis. In normal mice, the high levels of *ptc2* in the primary and secondary spermatocytes suggest that Sertoli-derived Dhh also signals to the developing sperm (Carpenter et al. 1998). Consistent with this idea, a mouse homolog of *fused* (*fu*), a cytosolic component of Hh signaling in *Drosophila*, is expressed in the primary and secondary spermatocytes along with *ptc2* (Carpenter et al. 1998). This indicates that spermatocytes are competent to respond to the Dhh. While it is not known how *ptc2* expression is affected in *Dhh* deficient males, the above evidence argues for a role of *ptc2* in regulating Dhh signaling during spermatogenesis.

In the hair follicle, as in the testis, *ptc1*, and *ptc2* have distinct patterns of expression. *Shh* is expressed in the epithelial cell layer of the developing hair follicle whereas *ptc1* is found in the surrounding mesenchyme (Motoyama et al. 1998). These expression patterns imply that Shh signals from the epithelia to adjacent mesenchymal cells. In contrast, *ptc2* is detected in the same epithelial cells that express *Shh* (Carpenter et al. 1998; Motoyama et al. 1998; Motoyama et al. 1998).

Both Ptc proteins appear to respond to Shh because Shh overexpression in the skin induces higher levels of both *ptc1* and *ptc2*, although the induction of *ptc2* is modest (Carpenter et al. 1998). Why are *ptc2* and *Shh* produced in the same cells? It may be that these epithelial cells are responding to Shh in an autocrine fashion or that *ptc2* levels are always high in the follicular epithelia to restrict or concentrate Shh at the site of secretion. Genetic studies in vertebrates may help elucidate the roles of *ptc2* in signaling.

12.3 Hh Signal Transmission from the Cytosol to the Nucleus

12.3.1 Cytosolic Components of Hh Signaling

The transduction of Hh signaling from the cell surface to the nucleus involves a host of unusual components and mechanisms. A significant number of these cytosolic components have been identified genetically in *Drosophila* (Table 1). In several cases, the genetic interactions have been substantiated with biochemical studies. How Hh signaling regulates many of these proteins is not well understood, but recent studies in *Drosophila* have identified several interactions critical for regulating transmission of the Hh signal to the nucleus.

The cytosolic proteins Costal2 (Cos2), Fused (Fu), and Cubitus interruptus (Ci) form part of a large complex that may regulate Hh-mediated gene expression (Robbins et al. 1997; Sisson et al. 1997). Cos2, a kinesin-like protein, binds microtubules and is believed to tether Ci, the major transcriptional regulator of Hh target genes, to the cytoskeleton and away from the nucleus (Robbins et al. 1997; Sisson et al. 1997). Fu, a putative serine-threonine kinase required for Hh signaling (Preat et al. 1990), associates with both Ci and Cos2. Activation of signaling is proposed to release Ci from this complex so that it can enter the nucleus and regulate gene expression (Robbins et al. 1997; Sisson et al. 1997). How this happens is not yet clear. Hh signaling may inhibit Cos2 function, activate the Fu kinase, or antagonize Suppressor of Fused – Su(fu) – a protein that associates with Fu and Ci to oppose Hh signaling (Pham et al. 1995; Monnier et al. 1998). The full nature of interactions

Table 1. Components of the Hh signaling pathway in *Drosophila*

Gene	Possible Function	Effect on Hh signaling
costal2[a]	Kinesin-related protein,	Negative
CREB binding protein[b]	Transcriptional coactivator	Positive
cubitus interruptus[c]	Zinc finger transcription factor	Positive
fused[d]	Serine/threonine kinase	Positive
hedgehog[e]	Secreted protein	Positive
oroshigane[f]	Unidentified	Positive
patched[g]	Membrane receptor	Negative
protein kinase A[h]	Serine/threonine kinase	Negative/Positive
slimb[i]	Ubiquitination pathway	Negative
smoothened[j]	Membrane protein	Positive
suppressor of fused[k]	PEST motif, antagonizes Ci	Negative
tout velu[l]	Facilitator of Hh diffusion	Positive

[a]Sisson et al. 1997; [b]Akimaru et al. 1997; [c]Orenic et al. 1990; [d]Preat et al. 1990); [e]Lee et al. 1992; Mohler and Vani 1992; Tabata et al. 1992; Tashiro et al. 1993); [f]Epps et al. 1997; [g]Hooper and Scott 1989; Nakano et al. 1989; [h]Jiang and Struhl 1995; Lepage et al. 1995; Li et al. 1995; Pan and Rubin 1995); [i]Jiang and Struhl 1998; Theodosiou et al. 1998); [j]Alcedo et al. 1996; van den Heuvel and Ingham 1996); [k]Pham et al. 1995; [l]Bellaiche et al. 1998).

between these and other cytosolic components and how they relay signal transmission is an area of active investigation.

12.3.2 Regulation of Ci Activity

In addition to controlling Ci entry into the nucleus, Hh signaling also prevents Ci proteolytic cleavage. Ci exists in at least two forms, the full-length 155 kDa protein and a 75 kDa cleavage product which functions as a transcriptional repressor (Aza-Blanc et al. 1997). In the absence of Hh, the 75 kDa repressor is generated constitutively. Activation of Hh signaling inhibits this cleavage and promotes the accumulation of full-length Ci. The exact nature of the activator form of Ci is nebulous since little or no full-length protein is ever detected in the nucleus (Motzny and Holmgren 1995; Aza-Blanc et al. 1997; Sisson et al. 1997). Recent work suggests that Hh signaling causes the activation or accumulation of a labile form of Ci that behaves as a transcriptional activator (Alves et al. 1998; Ohlmeyer and Kalderon 1998). Generation of this activated form of Ci appears to be mediated by Fu and Su(fu) but the mechanism by which these proteins do so is not known. The proteolytic cleavage of Ci may have arisen in *Drosophila* to permit the fast development of tissues. The fly embryo develops rapidly, with the larva emerging about 22 h after egg laying. A short period of development may require rapid control mechanisms, such as the posttranslational regulation of Ci, for turning developmental signals on and off at the appropriate time.

12.3.3 *Gli* Function in Vertebrates

Vertebrate homologs of Ci comprise the Gli family of zinc finger transcription factors, Gli1, Gli2, and Gli3 (Kinzler et al. 1988; Ruppert et al. 1988; Ruppert et al. 1990). Functional studies illustrate that Gli family members are clearly the mediators of vertebrate Hh signaling. In the mouse, *Gli1* is a target gene of Hh signaling; ectopic Shh expression induces ectopic *Gli1* (Grindley et al. 1997; Hynes et al. 1997) and in *ptc1* mutant mice, *Gli1* transcription is derepressed (Goodrich et al. 1997; Hahn et al. 1998). Like Ci in *Drosophila*, ectopic expression of Gli1 induces Hh targets such as *ptc1* (Hynes et al. 1997). In mice, *Shh* is essential for the proper development of the foregut and other organs (Litingtung et al. 1998). Likewise, mice deficient for *Gli2* and *Gli3* show defects which suggest that these transcription factors mediate Shh signaling in the foregut (Motoyama et al. 1998). In contrast to *Drosophila*, where Hh signaling posttranslationally regulates Ci protein, Gli family members appear to be regulated by Hh family members at the transcriptional level. Whether Gli family members undergo a proteolytic processing event like Ci is not known. The high degree of functional conservation in the Hh pathway demonstrated so far would argue for conservation of these regulatory mechanisms, but to date, posttranslational modification of Gli family members has yet to be established.

12.4 Hh Signaling in Disease and Tumorigenesis

12.4.1 *PTCH1*

The first indication of the importance of *ptc* in human disease came from the discovery of inactivating mutations in *PATCHED1* (*PTCH1*) in patients with the basal cell nevus syndrome (BCNS) (Hahn et al. 1996; Johnson et al. 1996). BCNS, also known as Gorlin's syndrome, is an autosomal-dominant disease associated with a wide variety of congenital defects and tumor types (Gorlin 1995). The defining characteristic in individuals with this syndrome is the appearance of large numbers of the skin tumor, basal cell carcinoma (BCC), during the second to third decade of life. BCCs also arise sporadically in normal individuals and are estimated to be the most common human cancer among individuals of northern European descent (Miller and Weinstock 1994). In contrast to BCNS, sporadic BCCs typically arise in low numbers and late in life. BCNS patients often have a number of developmental defects such as rib anomalies, spina bifida occulta, and cephalic abnormalities. A lower frequency of patients manifest polydactyly, cleft lip and/or palate, and occular abnormalities. It is postulated that the developmental defects of BCNS arise by mutation of one allele of *PTCH1* in the germ line. The skin tumors in BCNS individuals arise from somatic mutation of the second *PTCH1* allele resulting in the complete loss of PTCH1 function (Gailani et al. 1996; Hahn et al. 1996; Johnson et al. 1996). The involvement of *PTCH1* in BCNS demonstrates the critical function of this protein in both prenatal and postnatal life in humans.

Inherited cancer syndromes often have a variable spectrum of phenotypic abnormalities and in certain cases, such as *BRCA1* mutations, this variability correlates with the type of mutation (Gayther et al. 1995). BCNS has considerable variability in severity and presentation. Over 80% of BCNS mutations in *PTCH1* result in protein truncations at positions scattered throughout the protein (Gailani et al. 1996; Wicking et al. 1997; Aszterbaum et al. 1998). An examination of 28 separate *PTCH1* mutations revealed no apparent phenotype–genotype correlations; mutation of different *PTCH1* regions did not correlate with specific developmental defects (Wicking et al. 1997).

PTCH1 mutations have been identified in a number of sporadic tumors from otherwise normal individuals (Table 2). At least one-third

Table 2. Mutation of Hh signaling components in human sporadic tumors and developmental syndromes

Disease	Gene mutated
Basal cell carcinoma	*PTCH1*[a], *SMO*[b]
Primitive neuroectodermal tumor (medulloblastoma)	*PTCH1*[c], *SMO*[d]
Meningioma	*PTCH1*[e]
Transitional carcinoma of the bladder	*PTCH1*[f]
Trichoepithelioma	*PTCH1*[g]
Glioma	*GLI1*[h]
Basal cell nevus syndrome (Gorlin's syndrome)	*PTCH1*[i]
Holoprosencephaly	*SHH*[j]
Greig cephalopolysyndactyly	*GLI3*[k]
Pallister-Hall syndrome	*GLI3*[l]
Postaxial polydactyly type A	*GLI3*[m]
Multiple exostoses syndrome	*EXT1*[n]

[a]Gailani et al. 1996; Hahn et al. 1996; Johnson et al. 1996; [b]Reifenberger et al. 1998; Xie et al. 1998; [c]Pietsch et al. 1997; Raffel et al. 1997; Vorechovsky et al. 1997; Wolter et al. 1997; Xie et al. 1997; [d]Reifenberger et al. 1998; [e]Xie et al. 1997; [f]McGarvey et al. 1998; [g]Vorechovsky et al. 1997; [h]Kinzler et al. 1987; [i]Hahn et al. 1996; Johnson et al. 1996; [j]Belloni et al. 1996; Roessler et al. 1996; Roessler et al. 1997); [k]Vortkamp et al. 1991; Wild et al. 1997; [l]Kang et al. 1997; [m]Radhakrishna et al. 1997; [n]Ahn et al. 1995.

of sporadic BCCs have mutations in *PTCH1*, though this estimate is conservative because of the inefficiency of the mutation-screening techniques (Gailani et al. 1996). In BCCs where both copies of *PTCH1* have been examined, point mutations are found either in both alleles or in one allele accompanied by deletion of the second (Gailani et al. 1996). *PTCH1* mutations have been found in several types of sporadic primitive neuroectodermal tumors (PNET), including medulloblastoma. Medulloblastoma occurs in BCNS patients at a frequency of 1–2% (Evans et al. 1991), but in sporadic PNETs, *PTCH1* mutations have been found at a higher frequency, about 15% (Pietsch et al. 1997; Raffel et al. 1997; Vorechovsky et al. 1997; Wolter et al. 1997; Xie et al. 1997). *PTCH1* is also mutated in a low percentage of sporadic trichoepitheliomas (Vorechovsky et al. 1997), esophageal carcinomas (Maesawa et al. 1998), and transitional carcinomas of the bladder (McGarvey et al. 1998). In each of these tumors, the second allele of *PTCH1* is frequently deleted suggesting that tumor formation requires greatly reduced or complete loss of PTCH1 function.

Identification of *PTCH1* mutations in both sporadic and familial cases of basal cell carcinoma and medulloblastoma, while compelling in its consistency, is nonetheless a correlative argument. Targeted deletion of mouse *ptc1* and a number of transgenic experiments have provided convincing functional data that aberrant Hh signaling is a cause and not a consequence of tumorigenesis. Mice heterozygous for a *ptc1* deletion, like humans, suffer a variety of congenital malformations including polydactyly and rib anomalies (Goodrich et al. 1997; Hahn et al. 1998). Rhabdomyosarcomas and medulloblastomas, both seen in BCNS, are also seen in these mice at a relatively high frequency.

Interestingly, *ptc1* heterozygous mice do not develop basal cell carcinomas. However, mice overexpressing Shh in the skin do develop a condition strikingly similar to BCC in humans (Fan et al. 1997; Oro et al. 1997). Excess Shh ligand appears to be sufficient for tumorigenesis, presumably by saturating and inactivating Ptc proteins. Mice overexpressing a mutated form of Smo in the skin also develop a condition similar to BCC (Xie et al. 1998). These missense mutations, originally identified in sporadic BCCs, are proposed to make Smo resistant to Ptc inhibition and hence constitutively active. Why does unrestrained Shh or Smo activity in mouse skin give rise to BCC while inactivation of *ptc1* does not? Perhaps a second *ptc* gene, possibly *ptc2*, controls proliferation in the skin in parallel with, or in place of, *ptc1*. It is intriguing that this role of *ptc1* might not be conserved between mice and humans.

12.4.2 *GLI3*

Components at the distal end of the Hh signaling pathway, the *Gli* family of transcription factors, have also been implicated in human disorders. *Gli1* was originally identified as a putative oncogene that is amplified in human gliomas (Kinzler et al. 1987), and only later was linked to Hh signaling (Orenic et al. 1990). *Gli2* has important effects on the formation of the esophagus and trachea. *Gli2* deficient mice that are heterozygous for *Gli3,* develop tracheo-esophageal fistula (Motoyama et al. 1998). While this defect is a congenital abnormality of unknown etiology in humans, *Gli2* might be a candidate gene.

GLI3 has been directly implicated in three distinct human autosomal dominant disorders. Greig cephalopolysyndactyly (GCPS), Pallister-

Hall syndrome (PHS), and postaxial polydactyly type A (PAP-A) are all associated with variable digit defects and other developmental abnormalities. In each of these syndromes, nonsense mutations in *GLI3* have been identified that are predicted to delete specific regions of the protein. In individuals with GCPS, much of the protein is deleted, including the zinc finger domain that is proposed to mediate DNA binding (Vortkamp et al. 1991). In PHS patients, truncations occur more distally in the protein, preserving the zinc finger domain but removing a conserved domain of unknown function, the post-zinc finger (PZF) domain (Kang et al. 1997). Both PHS and GCPS are associated with craniofacial abnormalities in addition to polydactyly. In both cases the PZF domain is removed. PAP-A individuals have mutations in *GLI3* that are predicted to truncate the protein after the PZF domain (Radhakrishna et al. 1997). These individuals have polydactyly but no craniofacial defects, indicating that the PZF domain may be important for midline facial patterning. The region of GLI3 distal to the PZF domain, which is lost in all three syndromes, is believed to be necessary for normal digit formation.

Since *Gli3* mutations result in too many rather than too few digits, the normal role of Gli3 may be to antagonize Shh signaling in distal limb patterning. In support of this idea, Shh signaling during chick limb development inhibits *Gli3* transcription (Marigo et al. 1996). In addition, the mouse mutation, *extra toes* (*Xt*), which is associated with a *Gli3* mutation similar to GCPS, results in digit overgrowth (Hui and Joyner 1993). If *Gli3* opposes the proliferative effects of Shh in distal limb patterning, it is curious that *Gli3* mutations do not result in tumorigenesis as seen with *Ptc1*. Mouse models of these human diseases will help clarify the roles of *Gli* family members in regulating Hh signaling.

12.4.3 *SHH*

Hh family members are potent secreted molecules so it is not surprising that mutations in human *SHH* have been implicated in disease. Mice lacking *Shh* have profound patterning defects in the brain, spinal cord, axial skeleton, and the limbs (Chiang et al. 1996). These abnormalities closely resemble a group of birth defects in humans known as alobar holoprosencephaly (HPE). HPE arises when the forebrain fails to divide into right and left hemispheres. As in *Shh* mutant mice, alobar HPE in humans is often accompanied by cyclopia, a primitive nasal structure, and midline clefting. The striking similarity of *Shh* deficient mice to human HPE quickly led to the identification of *SHH* mutations in both familial and sporadic HPE (Roessler et al. 1996).

The majority of *SHH* mutations in HPE patients result in amino acid substitutions rather than premature stop codons (Roessler et al. 1996; Roessler et al. 1997). These mutations are predicted to reduce SHH function, in part by affecting a critical aspect of its biosynthesis. Normally, Hh proteins are synthesized as precursors that are cleaved by an autocatalytic reaction into N-terminal (Hh-N) and C-terminal fragments (Hh-C) (Lee et al. 1994; Bumcrot et al. 1995; Valentini et al. 1997). All of the signaling activity appears to reside in the highly conserved Hh-N (Porter et al. 1995). Hh-C, while devoid of signaling activity, possesses an enzymatic activity that is responsible for Hh self-cleavage (Lee et al. 1994; Porter et al. 1995). At least half of the HPE mutations occur in SHH-C and may prevent SHH from undergoing the cleavage reaction (Roessler et al. 1997). This is consistent with work in *Drosophila* where mutations in Hh-C impair Hh maturation and function (Porter et al. 1995). Recent evidence implicating cholesterol in Hh signaling sheds further light on the nature of these mutations.

12.5 The Role of Cholesterol in Hh Signaling

12.5.1 Attachment of Cholesterol to Hh

Cholesterol is important in two distinct aspects of Hh signaling: the generation of Hh protein and the transduction of the signal in receiving cells. Abrogation of either aspect has profound consequences for the developing embryo. Cholesterol is essential for the maturation of Hh. During biosynthesis, when SHh undergoes the autocatalytic cleavage, a cholesterol moiety is attached to Hh-N (Porter et al. 1996). The cholesterol modification makes Hh-N lipophilic and appears to tether it to the cell surface (Bumcrot et al. 1995; Porter et al. 1995). This anchoring event has important biological implications, as it appears to decrease the range of action of Hh while increasing Hh concentration at its site of synthesis. Expression in *Drosophila* of Hh-N lacking the cholesterol modification, results in marked pattern abnormalities that are not detected with the full-length protein (Porter et al. 1996). Without the cholesterol attachment, Hh-N diffuses farther and signals to cells that are normally beyond its range.

Vertebrates appear to have used this feature to vary the concentration of Shh-N during neural tube development. Shh is produced by the notochord to induce adjacent floor plate cells in a contact-dependent manner (Roelink et al. 1994) and at high concentrations (Roelink et al. 1995). The cholesterol attachment may anchor Shh-N to the notochord surface and increase its local concentration. Later in neural tube development, Shh-N produced by the floor plate induces motor neurons at a distance, but at lower protein concentrations (Roelink et al. 1995). How Shh signals over longer distances is not well understood. If the protein diffuses over many cell lengths, cleavage of the cholesterol tether might be required. To date, no such enzyme or activity has been reported.

12.5.2 Modulation of Shh Signal Reception by Sterols

Sterols also appear to modulate the response of target cells to Hh family members. Several lines of evidence suggest that inhibition of cholesterol synthesis or homeostasis causes cyclopia in animals, remarkably similar to that seen in *Shh* deficient mice. Exposure of pregnant ewes, mice, or hamsters to cyclopamine, a plant alkaloid that resembles cholesterol, induces fetal cyclopia and other developmental abnormalities reminiscent of Shh signaling defects (Keeler 1975; Keeler 1978). Patients with Smith-Lemli-Opitz Syndrome (SLOS), a human disorder with defects in delta 7-sterol reductase, the terminal enzyme of cholesterol synthesis, also manifest cyclopia as seen in individuals with HPE (Fitzky et al. 1998).

Recent studies have delimited what aspect of Shh signaling is disrupted by cyclopamine and inhibitors of distal cholesterol synthesis. Studies using neural and stomach explants indicate that these agents block the ability of target tissues to respond to Shh (Cooper et al. 1998; Incardona et al. 1998; Kim and Melton 1998). This effect can be reversed in some cases by the addition of exogenous cholesterol (Incardona et al. 1998). The agents do not appear to block Shh cleavage or the attachment of cholesterol (Cooper et al. 1998; Incardona et al. 1998). Neither do these drugs exert their effects by inducing a general cholesterol deficit since lovastatin and 25-hydroxycholesterol, potent inhibitors of the first steps in cholesterol biosynthesis, do not block Shh signaling (Cooper et al. 1998; Incardona et al. 1998). While the molecular target of these inhibitors is not yet known, a potential candidate may be Ptc.

12.5.3 Homology of Ptc to Sterol-Sensing Proteins

The drugs that inhibit Hh signal reception may be acting on Ptc through a hydrophobic region of the protein called a sterol-sensing domain. This domain is a sequence motif containing five potential transmembrane domains. Sterol sensors are found in proteins involved in cholesterol metabolism, such as HMG CoA reductase (Kumagai et al. 1995) and SREBP Cleavage Activating Protein (SCAP) (Nohturfft et al. 1998). The sterol sensor is required in these proteins for their regulation by sterols. In the case of HMG CoA reductase, sterols increase protein degradation (Jingami et al. 1987), while for SCAP, sterols block its ability to activate a transcriptional response (Nohturfft et al. 1996). The link of sterol-sensing domains and cholesterol metabolism has recently been extended to an additional protein called NPC1.

   NPC1 is implicated as the defective gene in Niemann–Pick Type C1 syndrome, (Carstea et al. 1997; Loftus et al. 1997). This rare recessive disorder, which affects both humans and mice, results in the abnormal accumulation of cholesterol in response to the LDL receptor pathway (Liscum and Faust 1987; Pentchev et al. 1987). *NPC1* encodes a large protein with multiple transmembrane domains and is proposed to move between intracellular compartments to traffic cholesterol or other components (Carstea et al. 1997; Loftus et al. 1997). NPC1 and Ptc share sequence similarity in many of the predicted membrane-spanning regions including the proposed sterol sensor. The similarity of Ptc to proteins that respond to or traffic sterols suggests new functions. Ptc activity may be modulated by sterols and Ptc may regulate signaling in part by moving between compartments within the cell.

   The effects of cholesterol synthesis inhibitors on Shh signaling might be explained by alterations in Ptc activity as mediated through the sterol sensor. Several missense mutations in human *PTCH1* that appear to abolish its function, map to the putative sterol sensor domain (Chidambaram et al. 1996; Wicking et al. 1997). This model would predict that inhibitors of Shh block signal transduction by keeping Ptc in an active state regardless of the presence of Shh. How this might happen mechanistically and whether the drugs would interact directly with Ptc are unresolved questions. Reception of the Hh signal may involve endocytosis of Hh, Ptc, and Smo and subsequent intracellular movement. This would be consistent with Ptc localization in *Drosophila* where the protein is found at the plasma membrane and in intracellular vesicles (Capdevila et al. 1994). Interestingly, the sterol sensor proteins HMG CoA reductase and SCAP are not found at the cell surface but instead localize to the endoplasmic reticulum. Whether Ptc is a target of the cholesterol inhibitors or not, it is clear that sterols have an important role in Hh signaling. Functional studies of the conserved regions between Ptc and NPC1, including the putative sterol sensor, should increase our understanding of how Ptc and other components regulate signaling to control development and proliferation.

12.6 Conclusions

hh was originally identified in *Drosophila* as a gene involved in early embryonic patterning (Nusslein-Volhard and Wieschaus 1980). Since its discovery, a number of pathway components have been found in a wide range of animals. As in flies, *hh* family members in vertebrates are crucial to the proper development of a number of tissues and organ systems. Recent work has implicated Hh signaling components in human diseases and tumorigenesis. Mutations in the ligand, Shh, the reception complex, Ptc and Smo, and the transcription factors, Gli1 and Gli3, have all been implicated as primary causes of human diseases. Furthermore, a growing body of evidence continues to extend the roles of sterols in Hh signaling, from Hh biosynthesis to the response of target tissues to Hh. The homology of Ptc to NPC1 and to proteins regulated by sterols has provided exciting possibilities on how Hh signaling might be controlled. Future studies on Hh signal reception, how signaling is relayed from membrane to cytoplasm, and how sterols modulate these events, will greatly enhance our understanding of Hh signaling. This pathway has provided, and will continue to provide, valuable insights into the basic processes of development and disease.

References

Ahn J, Ludecke HJ, Lindow S, Horton WA, Lee B, Wagner MJ, Horsthemke B, Wells DE (1995) Cloning of the putative tumour suppressor gene for hereditary multiple exostoses (EXT1). Nat Genet 11:137–143

Akimaru H, Chen Y, Dai P, Hou DX, Nonaka M, Smolik SM, Armstrong S, Goodman RH, Ishii S (1997) *Drosophila* CBP is a co-activator of cubitus interruptus in hedgehog signalling. Nature 386:735–738

Alcedo J, Ayzenzon M, Von Ohlen T, Noll M, Hooper JE (1996) The *Drosophila* smoothened gene encodes a seven-pass membrane protein, a putative receptor for the hedgehog signal. Cell 86:221–232

Alves G, Limbourg-Bouchon B, Tricoire H, Brissard-Zahraoui J, Lamour-Isnard C, Busson D (1998) Modulation of hedgehog target gene expression by the fused serine- threonine kinase in wing imaginal discs. Mech Dev 78:17–31

Aszterbaum M, Rothman A, Johnson RL, Fisher M, Xie J, Bonifas JM, Zhang X, Scott MP, Epstein EH Jr (1998) Identification of mutations in the human PATCHED gene in sporadic basal cell carcinomas and in patients with the basal cell nevus syndrome. J Invest Dermatol 110:885–888

Aza-Blanc P, Ramirez-Weber FA, Laget MP, Schwartz C, Kornberg TB (1997) Proteolysis that is inhibited by hedgehog targets Cubitus interruptus protein to the nucleus and converts it to a repressor. Cell 89:1043–1053

Bellaiche Y, The I, Perrimon N (1998) Tout-velu is a *Drosophila* homologue of the putative tumour suppressor EXT-1 and is needed for Hh diffusion. Nature 394:85–88

Belloni E, Muenke M, Roessler E, Traverso G, Siegel-Bartelt J, Frumkin A, Mitchell HF, Donis-Keller H, Helms C, Hing AV, Heng HH, Koop B, Martindale D, Rommens JM, Tsui LC, Scherer SW (1996) Identification of Sonic hedgehog as a candidate gene responsible for holoprosencephaly. Nat Genet 14:353–356

Bitgood MJ, Shen L, McMahon AP (1996) Sertoli cell signaling by Desert hedgehog regulates the male germline. Curr Biol 6:298–304

Bumcrot DA, Takada R, McMahon AP (1995) Proteolytic processing yields two secreted forms of sonic hedgehog. Mol Cell Biol 15:2294–2303

Capdevila J, Pariente F, Sampedro J, Alonso JL, Guerrero I (1994) Subcellular localization of the segment polarity protein *patched* suggests an interaction with the *wingless* reception complex in *Drosophila* embryos. Development 120:987–998

Carpenter D, Stone DM, Brush J, Ryan A, Armanini M, Frantz G, Rosenthal A, de Sauvage FJ (1998) Characterization of two patched receptors for the vertebrate hedgehog protein family. Proc Natl Acad Sci USA 95:13630–13634

Carstea ED, Morris JA, Coleman KG, Loftus SK, Zhang D, Cummings C, Gu J, Rosenfeld MA, Pavan WJ, Krizman DB, Nagle J, Polymeropoulos MH, Sturley SL, Ioannou YA, Higgins ME, Comly M, Cooney A, Brown A, Kaneski CR, Blanchette-Mackie EJ, Dwyer NK, Neufeld EB, Chang TY, Liscum L, Strauss JF, Ohno K, Zeigler M, Carmi R, Sokol J, Markie D, O'Neill RR, van Diggelen OP, Elleder M, Patterson MC, Brady RO, Vanier MT, Pentchev PG, Tagle DA (1997) Niemann-Pick C1 disease gene: homology to mediators of cholesterol homeostasis. Science 277:228–231

Chen Y, Struhl G (1996) Dual roles for patched in sequestering and transducing Hedgehog. Cell 87:553–563

Chen Y, Struhl G (1998) In vivo evidence that Patched and Smoothened constitute distinct binding and transducing components of a Hedgehog receptor complex. Development 125:4943–4948

Chiang C, Litingtung Y, Lee E, Young KE, Corden JL, Westphal H, Beachy PA (1996) Cyclopia and defective axial patterning in mice lacking Sonic hedgehog gene function. Nature 383:407–413

Chidambaram A, Goldstein AM, Gailani MR, Gerrard B, Bale SJ, DiGiovanna JJ, Bale AE, Dean M (1996) Mutations in the human homologue of the *Drosophila* patched gene in Caucasian and African-American nevoid basal cell carcinoma syndrome patients. Cancer Res 56:4599–4601

Concordet JP, Lewis KE, Moore JW, Goodrich LV, Johnson RL, Scott MP, Ingham PW (1996) Spatial regulation of a zebrafish patched homologue reflects the roles of sonic hedgehog and protein kinase A in neural tube and somite patterning. Development 122:2835–2846

Cooper MK, Porter JA, Young KE, Beachy PA (1998) Teratogen-mediated inhibition of target tissue response to Shh signaling. Science 280:1603–1607

de Kretser DM, Loveland KL, Meinhardt A, Simorangkir D, Wreford N (1998) Spermatogenesis. Hum Reprod 13:1–8

Epps JL, Jones JB, Tanda S (1997) oroshigane, a new segment polarity gene of *Drosophila melanogaster*, functions in hedgehog signal transduction. Genetics 145:1041–1052

Evans DG, Farndon PA, Burnell LD, Gattamaneni HR, Birch JM (1991) The incidence of Gorlin syndrome in 173 consecutive cases of medulloblastoma. Br J Cancer 64:959–961

Fan H, Oro AE, Scott MP, Khavari PA (1997) Induction of basal cell carcinoma features in transgenic human skin expressing Sonic Hedgehog. Nat Med 3:788–792

Fitzky BU, Witsch-Baumgartner M, Erdel M, Lee JN, Paik YK, Glossmann H, Utermann G, Moebius FF (1998) Mutations in the Delta7-sterol reductase gene in patients with the Smith-Lemli-Opitz syndrome. Proc Natl Acad Sci USA 95:8181–8186

Gailani MR, Stahle-Backdahl M, Leffell DJ, Glynn M, Zaphiropoulos PG, Pressman C, Unden AB, Dean M, Brash DE, Bale AE, Toftgard R (1996) The role of the human homologue of *Drosophila* patched in sporadic basal cell carcinomas. Nat Genet 14:78–81

Gayther SA, Warren W, Mazoyer S, Russell PA, Harrington PA, Chiano M, Seal S, Hamoudi R, van Rensburg EJ, Dunning AM, et al. (1995) Germline mutations of the BRCA1 gene in breast and ovarian cancer families provide evidence for a genotype-phenotype correlation. Nat Genet 11:428–433

Goodrich LV, Johnson RL, Milenkovic L, McMahon JA, Scott MP (1996) Conservation of the hedgehog/patched signaling pathway from flies to mice: induction of a mouse patched gene by Hedgehog. Genes Dev 10:301–312

Goodrich LV, Milenkovic L, Higgins KM, Scott MP (1997) Altered neural cell fates and medulloblastoma in mouse patched mutants. Science 277:1109–1113

Gorlin RJ (1995) Nevoid basal cell carcinoma syndrome. Dermatol Clin 13:113–125

Grindley JC, Bellusci S, Perkins D, Hogan BL (1997) Evidence for the involvement of the Gli gene family in embryonic mouse lung development. Dev Biol 188:337–348

Griswold MD (1995) Interactions between germ cells and Sertoli cells in the testis. Biol Reprod 52:211–216

Hahn H, Wicking C, Zaphiropoulous PG, Gailani MR, Shanley S, Chidambaram A, Vorechovsky I, Holmberg E, Unden AB, Gillies S, Negus K, Smyth I, Pressman C, Leffell DJ, Gerrard B, Goldstein AM, Dean M, Toftgard R, Chenevix-Trench G, Wainwright B, Bale AE (1996) Mutations of the human homolog of *Drosophila* patched in the nevoid basal cell carcinoma syndrome. Cell 85:841–851

Hahn H, Wojnowski L, Zimmer AM, Hall J, Miller G, Zimmer A (1998) Rhabdomyosarcomas and radiation hypersensitivity in a mouse model of Gorlin syndrome. Nat Med 4:619–622

Hidalgo A (1991) Interactions between segment polarity genes and the generation of the segmental pattern in *Drosophila*. Mech Dev 35:77–87

Hidalgo A, Ingham P (1990) Cell patterning in the *Drosophila* segment: spatial regulation of the segment polarity gene *patched*. Development 110:291–301

Hooper JE, Scott MP (1989) The *Drosophila patched* gene encodes a putative membrane protein required for segmental patterning. Cell 59:751–765

Hui CC, Joyner AL (1993) A mouse model of Greig cephalopolysyndactyly syndrome: the extra-toes mutation contains an intragenic deletion of the Gli3 gene. Nat Genet 3:241–246

Hynes M, Stone DM, Dowd M, Pitts-Meek S, Goddard A, Gurney A, Rosenthal A (1997) Control of cell pattern in the neural tube by the zinc finger transcription factor and oncogene Gli-1. Neuron 19:15–26

Incardona JP, Gaffield W, Kapur RP, Roelink H (1998) The teratogenic Vera-trum alkaloid cyclopamine inhibits sonic hedgehog signal transduction. De-velopment 125:3553–3562

Ingham PW, Taylor AM, Nakano Y (1991) Role of the *Drosophila patched* gene in positional signalling. Nature 353:184–187

Jiang J, Struhl G (1995) Protein kinase A and hedgehog signaling in *Drosophila* limb development. Cell 80:563–572

Jiang J, Struhl G (1998) Regulation of the Hedgehog and Wingless signalling pathways by the F- box/WD40-repeat protein Slimb. Nature 391:493–496

Jingami H, Brown MS, Goldstein JL, Anderson RG, Luskey KL (1987) Partial deletion of membrane-bound domain of 3-hydroxy-3-methylglutaryl coen-zyme A reductase eliminates sterol-enhanced degradation and prevents for-mation of crystalloid endoplasmic reticulum. J Cell Biol 104:1693–1704

Johnson RL, Rothman AL, Xie J, Goodrich LV, Bare JW, Bonifas JM, Quinn AG, Myers RM, Cox DR, Epstein E Jr, Scott MP (1996) Human homolog of patched, a candidate gene for the basal cell nevus syndrome. Science 272:1668–1671

Kang S, Graham J, Jr., Olney AH, Biesecker LG (1997) GLI3 frameshift mutations cause autosomal dominant Pallister-Hall syndrome. Nat Genet 15:266–268

Keeler RF (1975) Teratogenic effects of cyclopamine and jervine in rats, mice and hamsters. Proc Soc Exp Biol Med 149:302–306

Keeler RF (1978) Cyclopamine and related steroidal alkaloid teratogens: their occurrence, structural relationship, and biologic effects. Lipids 13:708–715

Kim SK, Melton DA (1998) Pancreas development is promoted by cy-clopamine, a hedgehog signaling inhibitor. Proc Natl Acad Sci USA 95:13036–13041

Kinzler KW, Bigner SH, Bigner DD, Trent JM, Law ML, O'Brien SJ, Wong AJ, Vogelstein B (1987) Identification of an amplified, highly expressed gene in a human glioma. Science 236:70–73

Kinzler KW, Ruppert JM, Bigner SH, Vogelstein B (1988) The GLI gene is a member of the Kruppel family of zinc finger proteins. Nature 332:371–374

Kumagai H, Chun KT, Simoni RD (1995) Molecular dissection of the role of the membrane domain in the regulated degradation of 3-hydroxy-3-methyl-glutaryl coenzyme A reductase. J Biol Chem 270:19107–19113

Lee JJ, von Kessler DP, Parks S, Beachy PA (1992) Secretion and localized transcription suggest a role in positional signaling for products of the seg-mentation gene hedgehog. Cell 71:33–50

Lee JJ, Ekker SC, von Kessler DP, Porter JA, Sun BI, Beachy PA (1994) Auto-proteolysis in hedgehog protein biogenesis. Science 266:1528–1537

Lepage T, Cohen SM, Diaz-Benjumea FJ, Parkhurst SM (1995) Signal transduction by cAMP-dependent protein kinase A in *Drosophila* limb pat-terning. Nature 373:711–715

Li W, Ohlmeyer JT, Lane ME, Kalderon D (1995) Function of protein kinase A in hedgehog signal transduction and *Drosophila* imaginal disc development. Cell 80:553–562

Liscum L, Faust JR (1987) Low density lipoprotein (LDL)-mediated suppression of cholesterol synthesis and LDL uptake is defective in Niemann-Pick type C fibroblasts. J Biol Chem 262:17002–17008

Litingtung Y, Lei L, Westphal H, Chiang C (1998) Sonic hedgehog is essential to foregut development. Nat Genet 20:58–61

Loftus SK, Morris JA, Carstea ED, Gu JZ, Cummings C, Brown A, Ellison J, Ohno K, Rosenfeld MA, Tagle DA, Pentchev PG, Pavan WJ (1997) Murine model of Niemann-Pick C disease: mutation in a cholesterol homeostasis gene. Science 277:232–235

Maesawa C, Tamura G, Iwaya T, Ogasawara S, Ishida K, Sato N, Nishizuka S, Suzuki Y, Ikeda K, Aoki K, Saito K, Satodate R (1998) Mutations in the human homologue of the *Drosophila* patched gene in esophageal squamous cell carcinoma. Genes Chromosomes Cancer 21:276–279

Marigo V, Tabin CJ (1996) Regulation of patched by Sonic hedgehog in the developing neural tube. Proc Natl Acad Sci USA 93:9346–9351

Marigo V, Davey RA, Zuo Y, Cunningham JM, Tabin CJ (1996) Biochemical evidence that patched is the Hedgehog receptor. Nature 384:176–179

Marigo V, Johnson RL, Vortkamp A, Tabin CJ (1996) Sonic hedgehog differentially regulates expression of GLI and GLI3 during limb development. Dev Biol 180:273–283

McGarvey TW, Maruta Y, Tomaszewski JE, Linnenbach AJ, Malkowicz SB (1998) PTCH gene mutations in invasive transitional cell carcinoma of the bladder. Oncogene 17:1167–1172

Miller DL, Weinstock MA (1994) Nonmelanoma skin cancer in the United States: incidence. J Am Acad Dermatol 30:774–778

Mohler J, Vani K (1992) Molecular organization and embryonic expression of the hedgehog gene involved in cell–cell communication in segmental patterning of *Drosophila*. Development 115:957–971

Monnier V, Dussillol F, Alves G, Lamour-Isnard C, Plessis A (1998) Suppressor of fused links fused and Cubitus interruptus on the hedgehog signalling pathway. Curr Biol 8:583–586

Motoyama J, Heng H, Crackower MA, Takabatake T, Takeshima K, Tsui LC, Hui C (1998) Overlapping and non-overlapping ptch2 expression with shh during mouse embryogenesis. Mech Dev 78:81–84

Motoyama J, Liu J, Mo R, Ding Q, Post M, Hui CC (1998) Essential function of Gli2 and Gli3 in the formation of lung, trachea and oesophagus. Nat Genet 20:54–57

Motoyama J, Takabatake T, Takeshima K, Hui C (1998) Ptch2, a second mouse Patched gene is co-expressed with Sonic hedgehog. Nat Genet 18:104–106

Motzny CK, Holmgren R (1995) The *Drosophila* Cubitus interruptus protein and its role in the wingless and hedgehog signal transduction pathways. Mech Dev 52:137–150

Nakano Y, Guerrero I, Hidalgo A, Taylor A, Whittle JR, Ingham PW (1989) A protein with several possible membrane-spanning domains encoded by the *Drosophila* segment polarity gene *patched*. Nature 341:508–513

Nohturfft A, Hua X, Brown MS, Goldstein JL (1996) Recurrent G-to-A substitution in a single codon of SREBP cleavage- activating protein causes sterol resistance in three mutant Chinese hamster ovary cell lines. Proc Natl Acad Sci USA 93:13709–13714

Nohturfft A, Brown MS, Goldstein JL (1998) Topology of SREBP cleavage-activating protein, a polytopic membrane protein with a sterol-sensing domain. J Biol Chem 273:17243–17250

Nusslein-Volhard C, Wieschaus E (1980) Mutations affecting segment number and polarity in *Drosophila*. Nature 287:795–801

Ohlmeyer JT, Kalderon D (1998) Hedgehog stimulates maturation of Cubitus interruptus into a labile transcriptional activator. Nature 396:749–753

Orenic TV, Slusarski DC, Kroll KL, Holmgren RA (1990) Cloning and characterization of the segment polarity gene Cubitus interruptus Dominant of *Drosophila*. Genes Dev 4:1053–1067

Oro AE, Higgins KM, Hu Z, Bonifas JM, Epstein E, Jr., Scott MP (1997) Basal cell carcinomas in mice overexpressing sonic hedgehog. Science 276:817–821

Pan D, Rubin GM (1995) cAMP-dependent protein kinase and hedgehog act antagonistically in regulating decapentaplegic transcription in *Drosophila* imaginal discs. Cell 80:543–552

Pentchev PG, Comly ME, Kruth HS, Tokoro T, Butler J, Sokol J, Filling-Katz M, Quirk JM, Marshall DC, Patel S (1987) Group C Niemann-Pick disease: faulty regulation of low-density lipoprotein uptake and cholesterol storage in cultured fibroblasts. FASEB J 1:40–45

Pham A, Therond P, Alves G, Tournier FB, Busson D, Lamour-Isnard C, Bouchon BL, Preat T, Tricoire H (1995) The Suppressor of fused gene encodes a novel PEST protein involved in *Drosophila* segment polarity establishment. Genetics 140:587–598

Pietsch T, Waha A, Koch A, Kraus J, Albrecht S, Tonn J, Sorensen N, Berthold F, Henk B, Schmandt N, Wolf HK, von Deimling A, Wainwright B, Chenevix-Trench G, Wiestler OD, Wicking C (1997) Medulloblastomas of the desmoplastic variant carry mutations of the human homologue of *Drosophila patched*. Cancer Res 57:2085–2088

Porter JA, von Kessler DP, Ekker SC, Young KE, Lee JJ, Moses K, Beachy PA (1995) The product of hedgehog autoproteolytic cleavage active in local and long-range signalling. Nature 374:363–366

Porter JA, Young KE, Beachy PA (1996) Cholesterol modification of hedgehog signaling proteins in animal development. Science 274:255–259

Preat T, Therond P, Lamour-Isnard C, Limbourg-Bouchon B, Tricoire H, Erk I, Mariol MC, Busson D (1990) A putative serine/threonine protein kinase encoded by the segment-polarity fused gene of *Drosophila*. Nature 347:87–89

Radhakrishna U, Wild A, Grzeschik KH, Antonarakis SE (1997) Mutation in GLI3 in postaxial polydactyly type A. Nat Genet 17:269–271

Raffel C, Jenkins RB, Frederick L, Hebrink D, Alderete B, Fults DW, James CD (1997) Sporadic medulloblastomas contain PTCH mutations. Cancer Res 57:842–845

Reifenberger J, Wolter M, Weber RG, Megahed M, Ruzicka T, Lichter P, Reifenberger G (1998) Missense mutations in SMOH in sporadic basal cell carcinomas of the skin and primitive neuroectodermal tumors of the central nervous system. Cancer Res 58:1798–1803

Robbins DJ, Nybakken KE, Kobayashi R, Sisson JC, Bishop JM, Therond PP (1997) Hedgehog elicits signal transduction by means of a large complex containing the kinesin-related protein costal2. Cell 90:225–234

Roelink H, Augsburger A, Heemskerk J, Korzh V, Norlin S, Ruiz i Altaba A, Tanabe Y, Placzek M, Edlund T, Jessell TM (1994) Floor plate and motor neuron induction by vhh-1, a vertebrate homolog of hedgehog expressed by the notochord. Cell 76:761–775

Roelink H, Porter JA, Chiang C, Tanabe Y, Chang DT, Beachy PA, Jessell TM (1995) Floor plate and motor neuron induction by different concentrations of the amino-terminal cleavage product of sonic hedgehog autoproteolysis. Cell 81:445–455

Roessler E, Belloni E, Gaudenz K, Jay P, Berta P, Scherer SW, Tsui LC, Muenke M (1996) Mutations in the human Sonic Hedgehog gene cause holoprosencephaly. Nat Genet 14:357–360

Roessler E, Belloni E, Gaudenz K, Vargas F, Scherer SW, Tsui LC, Muenke M (1997) Mutations in the C-terminal domain of Sonic Hedgehog cause holoprosencephaly. Hum Mol Genet 6:1847–1853

Ruppert JM, Kinzler KW, Wong AJ, Bigner SH, Kao FT, Law ML, Seuanez HN, O'Brien SJ, Vogelstein B (1988) The GLI-Kruppel family of human genes. Mol Cell Biol 8:3104–3113

Ruppert JM, Vogelstein B, Arheden K, Kinzler KW (1990) GLI3 encodes a 190-kilodalton protein with multiple regions of GLI similarity. Mol Cell Biol 10:5408–5415

Sisson JC, Ho KS, Suyama K, Scott MP (1997) Costal2, a novel kinesin-related protein in the Hedgehog signaling pathway. Cell 90:235–245

Stone DM, Hynes M, Armanini M, Swanson TA, Gu Q, Johnson RL, Scott MP, Pennica D, Goddard A, Phillips H, Noll M, Hooper JE, de Sauvage F,

Rosenthal A (1996) The tumour-suppressor gene patched encodes a candidate receptor for Sonic hedgehog. Nature 384:129–134

Tabata T, Eaton S, Kornberg TB (1992) The *Drosophila* hedgehog gene is expressed specifically in posterior compartment cells and is a target of engrailed regulation. Genes Dev 6:2635–2645

Takabatake T, Ogawa M, Takahashi TC, Mizuno M, Okamoto M, Takeshima K (1997) Hedgehog and patched gene expression in adult ocular tissues. FEBS Lett 410:485–489

Tashiro S, Michiue T, Higashijima S, Zenno S, Ishimaru S, Takahashi F, Orihara M, Kojima T, Saigo K (1993) Structure and expression of hedgehog, a *Drosophila* segment-polarity gene required for cell–cell communication. Gene 124:183–189

Taylor AM, Nakano Y, Mohler J, Ingham PW (1993) Contrasting distributions of patched and hedgehog proteins in the *Drosophila* embryo. Mech Dev 42:89–96

Theodosiou NA, Zhang S, Wang WY, Xu T (1998) slimb coordinates wg and dpp expression in the dorsal-ventral and anterior-posterior axes during limb development. Development 125:3411–3416

Valentini RP, Brookhiser WT, Park J, Yang T, Briggs J, Dressler G, Holzman LB (1997) Post-translational processing and renal expression of mouse Indian hedgehog. J Biol Chem 272:8466–8473

van den Heuvel M, Ingham PW (1996) smoothened encodes a receptor-like serpentine protein required for hedgehog signalling. Nature 382:547–551

Vorechovsky I, Tingby O, Hartman M, Stromberg B, Nister M, Collins VP, Toftgard R (1997) Somatic mutations in the human homologue of *Drosophila* patched in primitive neuroectodermal tumours. Oncogene 15:361–366

Vorechovsky I, Unden AB, Sandstedt B, Toftgard R, Stahle-Backdahl M (1997) Trichoepitheliomas contain somatic mutations in the overexpressed PTCH gene: support for a gatekeeper mechanism in skin tumorigenesis. Cancer Res 57:4677–4681

Vortkamp A, Gessler M, Grzeschik KH (1991) GLI3 zinc-finger gene interrupted by translocations in Greig syndrome families. Nature 352:539–540

Vortkamp A, Lee K, Lanske B, Segre GV, Kronenberg HM, Tabin CJ (1996) Regulation of rate of cartilage differentiation by Indian hedgehog and PTH-related protein. Science 273:613–622

Wicking C, Shanley S, Smyth I, Gillies S, Negus K, Graham S, Suthers G, Haites N, Edwards M, Wainwright B, Chenevix-Trench G (1997) Most germ-line mutations in the nevoid basal cell carcinoma syndrome lead to a premature termination of the PATCHED protein, and no genotype-phenotype correlations are evident. Am J Hum Genet 60:21–26

Wild A, Kalff-Suske M, Vortkamp A, Bornholdt D, Konig R, Grzeschik KH (1997) Point mutations in human GLI3 cause Greig syndrome. Hum Mol Genet 6:1979–1984

Wolter M, Reifenberger J, Sommer C, Ruzicka T, Reifenberger G (1997) Mutations in the human homologue of the *Drosophila* segment polarity gene patched (PTCH) in sporadic basal cell carcinomas of the skin and primitive neuroectodermal tumors of the central nervous system. Cancer Res 57:2581–2585

Xie J, Johnson RL, Zhang X, Bare JW, Waldman FM, Cogen PH, Menon AG, Warren RS, Chen LC, Scott MP, Epstein E, Jr. (1997) Mutations of the PATCHED gene in several types of sporadic extracutaneous tumors. Cancer Res 57:2369–2372

Xie J, Murone M, Luoh SM, Ryan A, Gu Q, Zhang C, Bonifas JM, Lam CW, Hynes M, Goddard A, Rosenthal A, Epstein E Jr, de Sauvage FJ (1998) Activating Smoothened mutations in sporadic basal-cell carcinoma. Nature 391:90–92

Subject Index